Lecture Notes in Artificial Intelligence 5680

Edited by R. Goebel, J. Siekmann, and W. Wahlster

Subseries of Lecture Notes in Computer Science

Longbing Cao   Vladimir Gorodetsky
Jiming Liu   Gerhard Weiss   Philip S. Yu (Eds.)

Agents and Data Mining Interaction

4th International Workshop, ADMI 2009
Budapest, Hungary, May 10-15, 2009
Revised Selected Papers

 Springer

Series Editors

Randy Goebel
University of Alberta, Edmonton, Canada
Jörg Siekmann
University of Saarland, Saarbrücken, Germany
Wolfgang Wahlster
DFKI and University of Saarland, Saarbrücken, Germany

Volume Editors

Longbing Cao
University of Technology, Sydney, Australia
E-mail: lbcao@it.uts.edu.au

Vladimir Gorodetsky
St. Petersburg Intitute for Informatics and Automation, St. Petersburg, Russia
E-mail: gor@mail.iias.spb.su

Jiming Liu
Hong Kong Baptist University, Kowloon Tong, Hong Kong
E-mail: jiming@comp.hkbu.edu.hk

Gerhard Weiss
Software Competence Center Hagenberg GmbH, Hagenberg, Austria
E-mail: gerhard.weiss@scch.at

Philip S. Yu
University of Illinois at Chicago, Chicago, IL 60607, USA
E-mail: psyu@cs.uic.edu

Library of Congress Control Number: 2009931690

CR Subject Classification (1998): I.2.11, I.2, H.2.8, H.3.3, C.2.4, D.2, F.2

LNCS Sublibrary: SL 7 – Artificial Intelligence

ISSN   0302-9743
ISBN-10   3-642-03602-3 Springer Berlin Heidelberg New York
ISBN-13   978-3-642-03602-6 Springer Berlin Heidelberg New York

springer.com

© Springer-Verlag Berlin Heidelberg 2009
Printed in Germany

Typesetting: Camera-ready by author, data conversion by Scientific Publishing Services, Chennai, India
Printed on acid-free paper   SPIN: 12723744   06/3180   5 4 3 2 1 0

Preface

The 2009 International Workshop on Agents and Data Mining Interaction (ADMI 2009) was a joint event with AAMAS2009.

In recent years, agents and data mining interaction (ADMI), or agent mining for short, has emerged as a very promising research field. Following the success of ADMI 2006 in Hong Kong, ADMI 2007 in San Jose, and ADMI 2008 in Sydney, the ADMI 2009 workshop in Budapest provided a premier forum for sharing research and engineering results, as well as potential challenges and prospects encountered in the synergy between agents and data mining.

As usual, the ADMI workshop encouraged and promoted theoretical and applied research and development, which aims at:

- Exploiting agent-driven data mining and demonstrating how intelligent agent technology can contribute to critical data mining problems in theory and practice
- Improving data mining-driven agents and showing how data mining can strengthen agent intelligence in research and practical applications
- Exploring the integration of agents and data mining toward a super-intelligent information processing and systems
- Identifying challenges and directions for future research on the synergy between agents and data mining

ADMI 2009 featured two invited talks and twelve selected papers. The first invited talk was on "Agents and Data Mining in Bioinformatics," with the second focusing on "Knowledge-Based Reinforcement Learning." The ten accepted papers are from seven countries. A majority of submissions came from European countries, indicating the boom of ADMI research in Europe. In addition the two invited papers, addressed fundamental issues related to agent-driven data mining, data mining-driven agents, and agent mining applications.

The proceedings of the ADMI workshops will be published as part of the LNAI series by Springer. We appreciate the support of Springer, and in particular Alfred Hofmann.

ADMI 2009 was sponsored by the Agent-Mining Interaction and Integration Special Interest Group (AMII-SIG: www.agentmining.org). We are grateful to the Steering Committee for their guidance.

More information about ADMI 2009 can be found on the workshop website: http://admi09.agentmining.org.

Finally, we appreciate the contributions made by all authors, Program Committee members, invited speakers, panelists, and the AAMAS2009 workshop and local organizers.

May 2009

Philip S. Yu
Longbing Cao
Vladimir Gorodetsky
Jiming Liu
Gerhard Weiss

Organization

General Chair

Philip S. Yu — Dept. of Computer Science, University of Illinois at Chicago, USA

Workshop Co-chairs

Longbing Cao — Faculty of Engineering and Information Technology, University of Technology, Sydney, Australia

Vladimir Gorodetsky — St. Petersburg Institute for Informatics and Automation, Russian Academy of Sciences, Russia

Jiming Liu — Department of Computer Science, Hong Kong Baptist University, China

Gerhard Weiss — Software Competence Center Hagenberg GmbH, Austria

Organizing Committee Co-chairs

Andreas Symeonidis — Aristotle University of Thessaloniki, Greece

Oleg Karsaev — St. Petersburg Institute for Informatics and Automation

Steering Committee

Longbing Cao — University of Technology Sydney, Australia (Coordinator)

Edmund H. Durfee — University of Michigan, USA

Vladimir Gorodetsky — St. Petersburg Institute for Informatics and Automation, Russia

Alfred Hofmann — Springer, Germany

Hillol Kargupta — University of Maryland Baltimore County, USA

Jiming Liu — Hong Kong Baptist University, China

Matthias Klusch — DFKI, Germany

Michael Luck — King's College London, UK

Pericles A. Mitkas — Aristotle University of Thessaloniki, Greece

Joerg Mueller — Technische Universität Clausthal, Germany

Ngoc Thanh Nguyen — Wroclaw University of Technology, Poland

Gerhard Weiss   Software Competence Center Hagenberg,
                                   Austria
Xindong Wu   University of Vermont, USA
Philip S. Yu   University of Illinois at Chicago, USA
Chengqi Zhang   University of Technology Sydney, Australia

Webmaster

Peerapol Moemeng   Faculty of Engineering and Information
                                   Technology, University of Technology,
                                   Sydney, Australia

Program Committee

Ahmed Hambaba   San Jose State University, USA
Ajith Abraham   Norwegian University of Science and
                                   Technology, Norway
Andrzej Skowron   Institute of Decision Process Support, Poland
Boi Faltings   Artificial Intelligence Laboratory, EPFL,
                                   Switzerland
Chengqi Zhang   University of Technology, Sydney, Australia
Daniel Kudenko   University of York, UK
Daniel Zeng   Arizona University, USA
David Taniar   Monash University, Australia
Dionysis Kehagias   ITI-CERTH, Greece
Agnieszka Dardzinska   Bialystok Technical University, Poland
Eduardo Alonso   City University London, UK
Eugenio Oliveira   University of Porto, Portugal
Heikki Helin   TeliaSonera Finland Oyj, Finland
Hillol Kargupta   University of Maryland, USA
Joerg Mueller   Technische Universität Clausthal, Germany
John Debenham   University of Technology Sydney, Australia
Juan Carlos Cubero   University of Granada, Spain
Kazuhiro Kuwabara   Ritsumeikan University, Japan
Ken Kaneiwa   NICT, Japan
Leonid Perlovsky   AFRL/IFGA, USA
Matthias Klusch   DFKI, Germany
Mehmet Orgun   Macquarie University, Australia
Mengchu Zhou   New Jersey Institute of Technology, USA
Michal Pechoucek   Czech Technical University, Czech Republic
Mirsad Hadzikadic   University of North Carolina, Charlotte, USA
Nathan Griffiths   University of Warwick, UK
Ngoc Thanh Nguyen   Wroclaw University of Technology, Poland
Pericles A. Mitkas   Aristotle University of Thessaloniki, Greece
Ran Wolff   Haifa University, Israel
Seunghyun Im   University of Pittsburgh at Johnstown, USA

Sung-Bae Cho	Yonsei University, Korea
Sviatoslav Braynov	University of Illinois at Springfield, USA
Valerie Camps	Paul Sabatier University, France
Vijay Raghavan	University of Louisiana, USA
Wee Keong Ng	Nanyang Technological University, Singapore
Wen-Ran Zhang	Georgia Southern University, USA
William Cheung	Hong Kong Baptist University, Hong Kong
Yanqing Zhang	Georgia State University, USA
Yasufumi TAKAMA	Tokyo Metropolitan University, Japan
Yiyu Yao	University of Regina, Canada
Yves Demazeau	CNRS, France
Zbigniew Ras	University of North Carolina, USA
Zili Zhang	Deakin University, Australia

Table of Contents

IV   Agent Mining Applications

Part I

Invited Talks and Papers

Agents and Data Mining in Bioinformatics: Joining Data Gathering and Automatic Annotation with Classification and Distributed Clustering

Ana L.C. Bazzan

Instituto de Informática, Universidade Federal do Rio Grande do Sul
Caixa Postal 15064, 91.501-970, Porto Alegre, RS, Brazil
bazzan@inf.ufrgs.br

Abstract. Multiagent systems and data mining techniques are being frequently used in genome projects, especially regarding the annotation process (annotation pipeline). This paper discusses annotation-related problems where agent-based and/or distributed data mining has been successfully employed.

1 Introduction

In many genome projects one of the key stages is the annotation process (annotation pipeline), which involves the handling of a large amount of data. Annotation employs a high number of programs and scripts that can easily be automated, freeing the specialist to carry out more valuable tasks. However, there is still a need for automated and integrated tools to support the annotation pipeline. One possibility here is to use artificial intelligence. Hence the motivation for this work is the application of techniques from multiagent systems (MAS) and data mining (DM) to solve problems in bioinformatics. This meets the recent trend around the integration of agent technologies and DM. In [9] (see also references therein) for instance, the authors identify and discuss two main challenges concerning the integration of agents and DM, namely *DM driven agent learning* and *agent driven DM*.

Knowledge discovery and DM are concerned with the use of ML techniques for knowledge extraction from large volumes of data. A relatively recent trend is to combine DM and multiagent systems approaches in this domain (e.g. [5,16,13,19]). The use of more sophisticated techniques for these applications is important due to the large amount of unexplored knowledge in the current mass of biological data.

In this paper agents encapsulate different machine learning (ML) algorithms. Several techniques such as negotiation and swarm intelligence are employed to construct an integrated domain model. In other words, agents are responsible for applying different machine learning algorithms and/or using subsets of the data to be mined and are able to cooperate to discover knowledge from these subsets.

L. Cao et al. (Eds.): ADMI 2009, LNCS 5680, pp. 3–20, 2009.

This approach has shown potential for many applications and it is opening interesting research questions. An important one is how to integrate the discovered knowledge by agents into a globally coherent model.

The paper is organized as follows. Section 2 presents the main aspects of bioinformatics and computational biology, restricted to those that require DM techniques. It also briefly presents the techniques employed in this work, e.g. how DM, ML, and MAS techniques can be employed in bioinformatics (in particular for annotation). Section 3 briefly describes some databases widely used in bioinformatics. Five particular problems from bioinformatics where DM techniques can be used are discussed in Sections 4, 5, 6, 7, and 8. respectively. The last section presents some concluding remarks.

2   Background

Bioinformatics or computational biology is a research area concerned with the investigation of tools and techniques from computing to solve problems from biology, particularly molecular biology. The reader is referred to [1,20] for more details. Several ML-based algorithms have been proposed in the literature for tasks such as finding particular signals associated with gene expression (e.g. [10]).

Protein function annotation includes classification of protein sequences according to their biological function and is an important task in bioinformatics. Of course, ideally, manually curated databases are preferred over automated processing. Issues such as the quality of data and compliance with standards could then be carefully analyzed. However, this approach is expensive, if feasible in the first place. With the large number of protein sequences entering into databases, it is desirable to annotate new incoming sequences by means of an automated method. Approaches based on DM and ML have been largely used, which explore functional annotation by using learning methods including decision trees and instance-based learning [3,12,18], neural networks [21], and support vector machines [8].

The most common way of annotating the function of a protein is to use sequence similarity (homology). Methods such as Smith-Waterman algorithm and BLAST have been largely used for measuring sequence similarity. However, the function of a protein with very small homology is difficult to assess using these methods, even using PSI-blast due to the low significance of the first hit. The same problem may arise for homologous proteins of different functions if one is only recently discovered and the other is the only known protein of similar sequence. This way, it is desirable to explore methods that are not based on sequence similarity. Protein sequence motifs can be seen an alternative approach.

Motifs can reveal important clues to a protein's role even if it is not globally similar to any known protein. The motifs for most important regions such as catalytic sites, binding sites, and protein-protein interaction sites are conserved over taxonomic distances and evolutionary time than are the sequences of the proteins themselves. In general, a pattern of motifs is required to classify a protein into a certain family of proteins [6]. This is a very valuable source of information as it gives a good idea of the variation of functions within the family, and hence a good indicative of how much functional annotation can safely be transferred. In summary, motifs are highly discriminative features for predicting the function of a protein.

Regarding the use of agent-based technologies in bioinformatics, these are useful because the necessary data is distributed among several sources, it is a dynamic area, its content is heterogeneous, and most of the work can be done in a parallel way. Hence, information agents can integrate multiple distributed heterogeneous information sources. There are only a few multiagent projects in the domain of bioinformatics. Next, some of these efforts are briefly reviewed.

In [11] a prototype is described whose aim is to automate the annotation of a sequence of a virus. Their work is based on information gathering: search, filtering, integration, analysis, and presentation of the data to the user. It uses the author framework DECAF, a multiagent system toolkit. The system has four overlapping multiagent organizations. The first, Basic Sequence Annotation, aims at integrating remote gene sequence annotations from various sources with the gene sequences at the Local Knowledge Base Management Agent (LKBMA). The second, Query, allows complex queries on the LKBMAs via a web interface. The third, Functional Annotation, is responsible for collecting information needed to guess the function of a gene, specifically using the Gene Ontology.

GeneWeaver [7] is a multiagent system for managing the task of genome analysis. Since all processes of identifying genes and predicting function of proteins (despite being labor-intensive and requiring expert knowledge) are computer-based tasks, it is possible to automate them. In case of Geneweaver this has been done using a multiagent approach. GeneWeaver is formed by a community of agents that interact with each other, each performing some distinct task, in an effort to automate the processes involved in, for example, determining protein function. Agents in the system can be concerned with the management of the primary databases, performing sequence analyses with existing tools, or storing and presenting resulting information.

A third agent-based tool is the MASKS environment [19], whose aim is to improve symbolic learning through knowledge exchange. The motivation of this work is to mimic human interaction in order to reach better solutions. This aims at supporting a recent practice in DM which is the use of collaborative systems. Inductors are combined in a multiagent system with autonomy to improve individual models through knowledge sharing. These are necessary because even if DM is a powerful technique for knowledge extraction, none of the embedded algorithm is good in all possible domains. Each algorithm contains an explicit or implicit bias that leads it to prefer certain generalizations over others. Therefore different DM techniques applied to the same dataset hardly generate the same result. In general, combining inductors increases the accuracy by reducing the bias. This integration aims at overcoming limitations of individual techniques through hybridization or fusion of various techniques. MASKS groups different symbolic ML algorithms encapsulated in agents in order to classify data. Its goal is to improve symbolic learning through knowledge exchange.

Automated annotation and ML are combined in [18]. A ML approach to generate rules based on already annotated keywords of the SWISS-PROT database is described. Such rules can then be applied to non-annotated protein sequences. Details can be found in [18] as well as in Section 4. In short, the authors have developed a method to automate the process of annotation regarding keywords in the SWISS-PROT database. This algorithm works on training data (in this case, previously annotated keywords regarding proteins) and generate rules to classify new instances. The training data comprises mainly taxonomy entries, motifs and patterns (see next section). Given these attributes C4.5 derives a classification rule for a target class (in this case, the keyword). Since dealing with the whole data in SWISS-PROT at once would be prohibitive, the authors divided it in protein groups according to the InterPro classification. Rules were generated and a confidence factor for each was calculated based on the number of false and true positives, by performing a cross-validation, and by testing the rate of error in predicting keyword annotation over the TrEMBL database.

3   Bioinformatics Databases

In bioinformatics, electronic databases are necessary due to the explosion and distribution of data related to the various genome projects. Here, the databases related to the present paper are briefly presented. SWISS-PROT (Uniprot)[1] is a curated database which provides a high level of annotation of each sequence, including: a description of the function of a protein, its domain structure, post-translational modifications, variants, etc. TrEMBL is structurally analogous to SWISS-PROT but it is a computer-annotated supplement of SWISS-PROT. SWISS-PROT has also extensive links to other databases, such as the databases of motifs and enzymes. The first lines are of each entry are the protein ID and accession number (AC). For our purposes the "DE" (description), the "DR" (cross reference to other databases), and the "KW" (keywords) lines are the most important. DE shows the name of the protein and its code. DR shows how an entry relates to other databases, in particular to motifs databases. This is important because some of these cross references will be used as attributes in the rule induction process, as it will be detailed. The KW field gives several hints to experts as to what regards proteins functions and structure and is the main target of the automatic annotation discussed in sections 4 and 5.

Since there are many motifs' recognition methods to address different sequence analysis problems, different databases of motifs exist, including those that contain relatively short motifs (e.g., PROSITE[2]); groups of motifs referred to as fingerprints (e.g. PRINTS[3]); or sequence patterns, often based on position-specific scoring matrices or Hidden Markov Models generated from multiple sequence alignments (e.g. Pfam[4]).

---

[1] http://www.expasy.ch/sprot/

[2] http://www.expasy.ch/prosite

[3] http://umber.sbs.man.ac.uk/dbbrowser/PRINTS/

[4] http://www.sanger.ac.uk/Pfam/

Interpro[5] integrates the most commonly used motifs databases, providing a layer on the top of them. Thus, a unique, non-redundant characterization of a given protein family, domain or functional site is created. InterPro receives the signatures from the member databases. The motifs are grouped manually into families (there are some automatic methods for producing the matches and over-lap files, but the ultimate decision for grouping and for determining relationships comes from a biologist).

Enzyme databases contain data about enzyme nomenclature, classification, functional characterization, and cross-references to others database, such as SWISS-PROT. ENZYME[6] is a repository of information relative to the nomen-clature of enzymes, based on the recommendations of the IUBMB. Enzymes are classified using the so-called EC (Enzyme Commission) codes. A given class represents the function of an enzyme and is specified by a number composed by four digits (e.g. 1.1.1.1 for alcohol dehydrogenase). The first digit indicates the general type of chemical reaction catalyzed by the enzyme. Considering that a particular protein can have several enzymatic activities, it can be classified into more than one EC classes. These proteins are considered multi-functional. Moreover, SWISS-PROT entries are also linked to the ENZYME database. As said, the line "DE" in the SWISS-PROT entry contains the EC numbers as-sociated to the corresponding protein. For example, the protein with AC equal to Q9I3W8 has two ECs numbers associated – 2.7.1.116 ([Isocitrate dehydroge-nase (NADP+)] kinase) and 3.1.3.-. (Hydrolases acting on ester bonds) – that are described in the "DE" line. These ECs are links to the ENZYME database. Linking these two databases is an efficient way to obtain the list of proteins annotated with the EC classes. Other important enzyme database is BRENDA (BRaunschweig ENzyme DAtabase).

4 An Agent-Based Environment for Automating Annotation

In spite of the growing number of tools for analyzing DNA sequences, there is no easy solution to the annotation problem. The shortcomings are twofold. First, while no single program can outperform the human expert, when several different programs are used, the performance can increase. Second, each program has its own output format, thus posing difficulties when one wants to compare results. The aim of the Agent-based environmenT for aUtomatiC annotation of Genomes (ATUCG) [4] is to provide the team in charge of genome sequencing and annotation with a tool that facilitates the activities involved in carrying out those tasks.

In ATUCG agents are allocated to do most of the automated methods. Be-sides, agents are in charge of performing much of the tedious work required when using different programs and tools, namely the translation or formating of inputs

---

[5] http://www.ebi.ac.uk/interpro/
[6] http://au.expasy.org/enzyme/

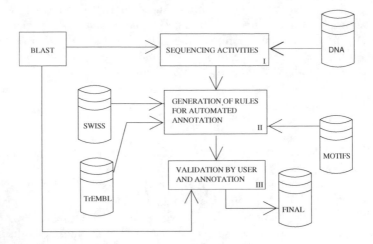

Fig. 1. Overview of the three–layer architecture

and outputs. ATUCG consists of three layers, each having several agents. The interrelationships and general overview are depicted in Figure 1.

Layer I aims at automating the tasks behind the process of finding open reading frames (ORFs). Having received a file containing the DNA sequencing of a given organism, layer I helps the user to define the ORFs. To this means, specific programs are used, each regarding the expertise of a particular agent. The output that layer I passes to layer II is a file containing a list of non-redundant ORFs. Layer II is associated with three main tasks: extraction and formatting of data, automatic annotation of data regarding profiles or families of proteins, and generation and validation of rules to automatically annotate the Keywords field in the SWISS-PROT database. Layer III permits the user to validate the automatic annotation by helping him/her to perform the verification of the proposed annotation, mainly by "translating" the annotation rules into a semantically richer language and presenting it in a more understandable form. This is done by linking attributes presented in the rules to more complete data extracted from databases such as PROSITE and InterPro. Once the annotations are verified, they are collected into two files: one for accepted annotations, another for rejected ones. The present text concentrates on layer II (next section), focusing on ML algorithms. The reader is referred to [4] for more details on the other layers.

The aim of layer II is threefold: design of agents that are experts on data extraction (e.g. from SWISS-PROT and InterPro databases) and formatting; creation of a database of rules for annotation (this involves training and validation stages using standard ML techniques); automatization of the annotation of fields related to motifs or profiles regarding families of proteins (i.e. annotation regarding PROSITE and InterPro attributes). The input to this layer is a file with a nonredundant list of ORFs. Besides, layer II has connections to several databases for in this layer training data has to be gathered in order to produce rules for

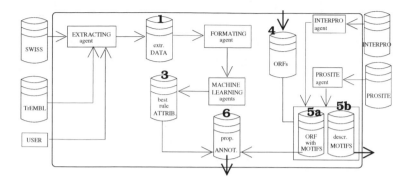

Fig. 2. Layer II

automatic annotation. The main output passed to layer III is a set of proposed annotation for selected attributes. Figure 2 depicts the various databases and agents involved in the tasks associated with the layer II. As already mentioned, three main tasks are carried out in this layer: data extraction and reformatting; application of different ML techniques to train data and produce the rules for annotation; and the annotation regarding protein profiles, keywords, and other attributes. The process called "data extraction and formating" deals with accessing two main databases (SWISS-PROT and TrEMBL) aiming to retrieve training and validation data respectively. The agent called "extraction agent" is invoked and has the following tasks: ask the user for attributes which will be used in the training process, retrieving data, and preparing the retrieved data in order to serve as input for our ML tools.

The attributes that the user typically wants to work with are proteins (coded according to the SWISS-PROT AC's) and attributes such as classification of proteins in domains or families. These classifications are stored in the databases of protein families: PROSITE, InterPro, PFAM, and eventually others. This procedure retrieves data regarding accession number of proteins, keywords, and classification regarding the protein families. Then the preprocessed data is sent to the various agents responsible for each ML algorithm. Steps 1 and 2 (asking the user and retrieving the data) are relatively simple. However, step 3 is not only more complex, but can also be performed in several alternative ways since each ML algorithm makes use of proper syntax to describe a hypothesis. Since the agents will eventually need to exchange their best rules for classification, it is necessary to have these transformed to the same format. The details of the training using ML algorithms, and the design and use of cooperative learning can be found in Bazzan *et al.* [3] and Schroeder and Bazzan [19] respectively.

The rules which will determine whether or not a given protein can be annotated with a given keyword are generated using ML algorithms. In all cases the algorithms follow approximately the same approach. First, a training set is given to the algorithm. Depending on this algorithm, the output varies but, as mentioned before, there is an agent which is in charge of translating all outputs to a standard format. Then, agents receive and process the data, and generate their annotation rules. Regarding the different learning elements, symbolic classification techniques similar to C4.5 are used: CN2, T2, and Ripper. Learning happens in two stages. The first one is dedicated to individual learning. The input is the pre-processed training set and the configuration set. After that, the agent applies its rules of classification to the examples. The objective of the individual learning is to create an individual model of the domain problem. This model is a set of rules that were approved by the "evaluation element".

As soon as the individual learning is over, the rules created for the classification are evaluated with the standard n-fold cross-validation. The "evaluation element" measures each rule quality by executing a given rule evaluation function and stores those that are equal or better than the threshold (set by the user) to the agent knowledge base. Most of the time the individual learning stage produces a rule set with well described target classes along with poor described ones. This happens due to the algorithm heuristics applied to the data to extract knowledge. In the next stage, cooperative learning, the goal is to improve the individual results. The input for the cooperation stage is the knowledge base that reflects the knowledge obtained during the individual learning. The cooperative learning consists of two further steps. During the first one, the agent queries other agents' knowledge bases. The first agent to start the interaction is the one with the poorest overall accuracy. The agent searches for its equivalent rules with better quality. The rules that fill this requisite are added to the agent "model merge". Each agent repeats this process from the poorer to the richer average accuracy. One says that a rule is equivalent to another when the two describe the same class and the attributes overlap. This way a high quality rule is added to the learner ModelMerge when it is similar, overlaps, subsumes, or is in conflict with a low quality rule produced by the learner. For example, consider two rules $R1$ and $R2$ that describe class **C**. Rule $R1$ contains the attribute-value test for attributes **x** and **y**, while rule $R2$ includes tests with attributes **x** and **z**. These two rules are related. When the communication ends, the individual knowledge that was not changed is copied into the "modelMerge" component. At this moment each agent possesses two distinct models of the domain problem.

When the cooperative learning ends, the new model stored in the agent ModelMerge component is evaluated by using the test file. The individual model resultant is the one that covered the highest number of instances from the test file. The output generated by the environment is the best agent model. At the end of this process one has a list of the best rules to describe how to annotate each attribute.

After data extraction, reformatting, and application of ML techniques to train data and produce the rules for annotation, the last activity in layer II is the annotation related to profiles of motifs in proteins. This is mandatory since the agent which actually performs the annotation automatically would not work without the necessary annotation of the profiles regarding the proteins just discovered (because they are unknown, there is no information concerning profiles of motifs). To accomplish this, an agent reads each ORF in file 4 (Figure 2) and query both the PROSITE agent and the InterPro agent. Basically these then call a script to send the ORF to the PROSITE and InterPro web sites in order to get a list of profiles which correspond to the specific ORF. Besides, these two agents have to divide the profile information and save it to two distinct databases (files 5a and 5b in Figure 2). File 5a contains the ORF and its classification (list of applicable PROSITE and InterPro codes) while file 5b contains not only the codes but also the rest of the description of the profile itself. It is possible to ask the user to validate the inclusion of the profiles in both files 5a and 5b. This extension in the system is planned and would be a further check point in the automatization.

Finally, the last and central activity in layer II is the annotation of the KW field. Having partially annotated the record for each protein (for instance, each already has information on profiles and taxonomy), the next step is to apply the rules generated by the ML algorithms. This is the task of the "keyword agent". It reads files 3 and 5a (Figure 2) and do the following: i) reads an entry from file 5a (ORF together with applicable profiles); ii) repeats for each rule in file 3 (remember that each rule corresponds to a different keyword): parse the rule; tests all preconditions and concludes whether or not the keyword should be annotated; do the annotation: creates an entry in file 6 for the ORF, copy the attributes such as profiles, and add the keyword (in case a keyword does not apply, insert nothing); iii) close files. At the end of this procedure, two files are sent to the next layer: files 5b and 6. In layer III, the user can verify the annotation.

5 Integrating Knowledge through Cooperative Negotiation

In the previous section the ATUCG framework was presented. The case in which the rules generated are in conflict is discussed next, based on a simple negotiation process. Negotiation is a process in which two or more parties make a joint decision. It is a key form of interaction that enables groups of agents to arrive to mutual agreement. Hence the basic idea behind negotiation is reaching a consensus. Cooperative negotiation is a particular kind of negotiation where agents cooperate and collaborate to achieve a common objective, in the best interest of the system as a whole. In cooperative negotiation, each agent has a partial view of the problem and the results are put together via negotiation trying to solve the conflicts that arise from the partial knowledge.

The negotiation process involves two types of agents: learning agents and the mediator agent. The learning agents encapsule the ML algorithms and the mediator agent is responsible for controlling the communication among the learning agents and finalize the negotiation. The negotiation process starts with the mediator agent asking the learning agents to send their overall accuracies. The first learning agent to generate a proposal is the one that has the poorest overall accuracy. The proposal contains the first rule in the rule set of this agent, which has not yet been evaluated, and which has an accuracy equal or better than a threshold. Rules that do not satisfy the threshold are not considered. This proposal is then sent to the mediator agent, which repasses it to others agents. Each agent then evaluates the proposal, searching for a equivalent rule with better accuracy. This evaluation requires comparison of rules and, therefore, it is necessary to establish equivalences for them. One rule is equivalent to another one when both describe the same concept and at least one attribute overlaps. If an equivalent rule is not found, or an equivalent one does not have better accuracy, then the agent accepts the proposal. Otherwise, if the agent has a rule with a better accuracy than the proposed rule, then its rule is sent as a counter-proposal to the mediator agent, which evaluates the several counter-proposals received. This is so because several agents can send a counter-proposal. The one which has the best accuracy is selected. The selected counter-proposal is then sent to the other agents (including the agent that generated the first proposal). These review it, and the process is repeated. When a proposal or a counter-proposal is accepted by all agents, the mediator adds the corresponding rule in the integrated model and the learning agents mark its equivalent one as evaluated. The negotiation ends when all rules were evaluated.

5.1   Case Study in Automated Annotation of Proteins

This negotiation approach was used in [13] to induce a model for annotation of proteins, regarding specifically the field KW in Swiss-Prot. Proteins from the model organism *Arabidopsis thaliana* were used, which are available in public databases such as Swiss-Prot. Using it, 3038 proteins were found that relates to *A. thaliana*. 347 keywords appeared in the data but only those whose number of instances was higher than 100 were considered. The number of keywords satisfying this criterion was 29. The attributes that appeared in this data are only the AC's for all attributes related to the Interpro, which appear in Swiss-Prot as a cross-referenced database. The number of attributes of this domain is 1496. For each keyword, the collected data was split in three subsets: training (80% – 2430 instances), validation (10% – 304 instances) and test (10% – 304 instances).

The training set was split in four subsets, each one with approximately 607 examples. Each subset of the training set was assigned to one of the learning agents A, B, C, and D. Besides, two agents encapsulate the C4.5 algorithm (agents A and C) and the other two (agents B and D) use the CN2 algorithm. These subsets were organized in input files, respecting the specific syntax of the C4.5 and CN2. Using the input data, the four learning agents induce rules for each keyword (target class).

Once the rules were generated, they were transformed into the PBM format. After the rule transformation process, the rules of the four agents are evaluated using the validation subset. For each rule of the model, its accuracy was estimated by applying the Laplace expected error measure. Only rules with accuracy equal or better than a threshold were considered. The overall accuracy of an agent is obtained from the average accuracy of its rules that satisfied the threshold. Once the evaluation of the individual models was made, the negotiation process starts. As said above, the negotiation process starts with the mediator agent asking to the four learning agents for their overall accuracies. The first learning agent to generate a proposal is the one that has the poorest overall accuracy. This then selects the first rule in its rule set that satisfies the threshold and sends it to the mediator agent, which repasses the rule to the others learning agents. These agents evaluate the rule, searching for an equivalent one, with better accuracy. If an equivalent rule with better accuracy than the proposed rule is not found, the agent that is evaluating the proposal accepts the proposed rule. Otherwise, if it has an equivalent rule with better accuracy, it sends it as a counter-proposal to the mediator agent, which evaluates the several rules received and selects the one that has the best accuracy. The selected rule is then sent to other agents which review it.

A rule is added to the integrated model if it is accepted by all agents. This rule and its equivalent one are marked as evaluated by the corresponding agent during the evaluating rules process. The negotiation process is repeated until there are no more rules to evaluate. This process was done for each keyword (i.e. the agents negotiate integrated models for each keyword).

5.2   Results and Discussion

The integrated model generated through cooperative negotiation was evaluated using the test subset. For comparative purposes, the individual models were also evaluated using this subset. Results showed that the integrated model had a better overall accuracy than the individual models regarding all keywords.

Also experiments with only two agents, each using all training data, were performed. In this case, each agent encapsulates one of the ML algorithms – C4.5 and CN2. Both used all 2430 instances of the training set to induce its model. The validation was performed using the instances of the validation set. The C4.5 agent obtained an overall accuracy (for all keywords) equal to 0.52 and the CN2 agent produced an accuracy of 0.50. For both agents, the quality of rules was poor when compared to the results obtained by the integrated model (0.89 – overall accuracy for all keywords). In one specific case, considering only the keyword "Iron", the C4.5 agent had an accuracy equal to 0.51 and CN2 agent presented an accuracy of 0.53, while the integrated model obtained an accuracy of 0.99. This happens because the amount of data is very high, thus the algorithms do not induce good models. Also, this task is time-consuming. These results have shown that the proposed approach can be applied to improve the accuracy of the individual models, integrating the better representations of each one.

6   Selection of Data Sets of Motifs as Attributes in the Process of Automating the Annotation of Proteins' Keywords

After proposing the DM-based approaches discussed in the previous sections, it was possible to notice that the quality of the annotation rules was poor when too many attributes were employed. Using all available data on protein families, patterns, and motifs as attributes for symbolic ML algorithms is prohibitive with symbolic machine learning methods. Thus the next step was to evaluate the information provided by those attributes, which can come from different data sets. Instead of using thousands of attributes during the ML process, [5] shows how to analyse which set of these attributes can potentially contribute more to the annotation process. Once those rules were generated, they were used to fill the KW field in the TrEMBL database.

Also in [5] the model organism *Arabidopsis thaliana* used to feed the Layer II of ATUCG. SWISS-PROT's cross-references were used as attributes in the ML process. Specifically PROSITE, Pfam, PRINTS, ProDom, and InterPro were used. Many keywords appeared in the data but the authors focused on those whose number of instances was higher than 100; the number of keywords satisfying this criterion was 27. Since the aim was to compare data sets of motifs, all motifs were used which are cross-referenced in SWISS-PROT as attributes. The number of attributes, by data set, was: 1316 (Intepro), 907 (Pfam), 220 (Prodom), 589 (Prosite), 246 (PRINTS), thus 3278 in total. Also, a constraint on the quality of the rules generated by C4.5 was imposed: each rule must cover a minimum number of 25 instances, a number that is approximately 1% of the number of training instances. The quality of each rule generated by C4.5 was evaluated via 5-fold cross-validation.

Table 1. Evaluation Test (5-fold CV) – Error, number of instances and confidence, for each data set of attributes (average over all keywords)

Database	Global Error (%)	Instances	Class (Keyword) Error (%)	Conf.	Non-Class Error (%)
All	37.98 (6.74)	49.50	31.94 (63.31)	0.62	0.65 (0.14)
Interpro+Prosite+Pfam	37.98 (6.74)	49.51	31.90 (63.19)	0.62	0.70 (0.15)
Interpro+Prosite	37.96 (6.73)	49.51	31.88 (63.16)	0.62	0.70 (0.15)
only Interpro	38.25 (6.79)	49.41	32.09 (63.51)	0.62	0.68 (0.15)
only Pfam	39.52 (7.01)	49.21	33.13 (65.38)	0.60	0.68 (0.15)
only Prosite	40.35 (7.15)	49.32	34.32 (68.92)	0.57	0.44 (0.09)
only Prints	43.70 (7.75)	48.64	37.13 (73.55)	0.47	0.31 (0.06)
only Prodom	52.54 (9.32)	50.36	47.77 (95.07)	0.20	0.23 (0.04)

Table 1 shows the error and confidence for each set of attributes (average over all keywords). When the classification is performed with attributes only from single databases in Table 1, in most cases the error in the non-class is low. However, looking at error rates regarding the positive class only (fourth column), some are unacceptable (e.g. 95.07% for ProDom). Similar conclusion can be drawn for confidence.

For the other data sets, the trend is that global error is low (e.g. 7.75% for PRINTS) but the error rate for the positive class is high. Better confidences and error rates are achieved when using the following databases: InterPro, Pfam, and also for the combinations: InterPro+PROSITE, and Inter-Pro+PROSITE+Pfam. However, in these last cases, the combination brought no increase: using attributes from InterPro alone is as good as using attributes from InterPro plus other data sets.

Finally, a note on the still high level of error rate. This is due to two main factors: low level of annotation of the KW field in SWISS-PROT and the unbalance of the two classes. These issues were investigated in [2].

7   Enzyme Classification

Another use of ATUCG is in classifying enzymes into functional classes according to the Enzyme Commission (EC) top level hierarchy. Such classifier will be then applied to not yet annotated proteins. This is indeed important because knowing to which family or subfamily an enzyme belongs may help in the determination of its catalytic mechanism and specificity, giving clues to the relevant biological function. Notice that this information is not directly available in databases such as Interpro. Although it may appear in SWISS-PROT (DE line), since it is manually curated, the EC classification is not always available. Thus the aim of the work described in [14] is exactly to fill this part of the DE line in an automatic way, for later insertion in TrEMBL.

Regarding training data, all proteins from the SWISS-PROT database were used in which the line "DE" is annotated with EC numbers. The first digit of the EC numbers appearing in this field in the corresponding SWISS-PROT entries were used. Using it, around 60,000 proteins annotated with EC numbers were found. The attributes that appear in this data are AC's related to InterPro domain classifications. These appear in SWISS-PROT as a cross-referenced database. This way, for each SWISS-PROT entry, the InterPro AC's and corresponding EC numbers, considering only the first and second digits, were extracted (i.e, each entry corresponds to one learning instance). These InterPro AC's were used as attributes by the learning algorithm. The attributes which appear in all examples related to one specific class are used to characterize it. The exact number of attributes found for each EC class are: 715 for oxidoreductase, 1269 for hydrolase, 231 for isomerase, 1309 for transferase, 391 for lyase, and 319 for ligase.

This data is organized in input files, one for each EC class, respecting the specific syntax of C4.5. Some enzymes are multi-functional and belong to more than one EC class. For these, more than one instance is created. For instance, the enzym with AC equal to Q9I3W8 is considered a positive instance to the EC 2 and EC 3 classes. This way, a total of 60488 instances were used.

Using the input data, C4.5 induces rules for a given target class. For example one rule basically suggests the annotation of the class "EC3" for a given enzyme if it belongs to the IPR010452 and if it does not belong to the IPR002381 families of proteins. The protein Q9I3W8 is correctly classified into the EC3 class, since it is related to the IPR010452.

Table 2, column 2, shows the results regarding the global classification error for each EC class, performing the 10-fold cross-validation procedure. The average error is 3.13% and 190 test instances had not been correctly classified (an average of 6048 test instances were used). The classes EC 5 and EC 6 which have the

Table 2. Results of the classification using 10-fold-cross-validation

EC	Global Error (%)	Class Instances	Class Error (%)	Class Conf.	Non Class Instances	Non Class Error (%)	Non Class Conf.
1	209.90 (3.50)	1160.3	198.00 (17.06)	0.98	4888.5	12.00 (0.25)	0.96
2	371.10 (6.10)	1958.4	342.20 (17.47)	0.98	4090.4	28.10 (0.69)	0.91
3	340.90 (5.60)	1485.7	305.10 (20.54)	0.96	4563.1	35.20 (0.77)	0.93
4	107.70 (1.80)	594.8	101.70 (17.10)	0.97	5454	5.80 (0.11)	0.98
5	60.90 (1.00)	283.6	54.20 (19.11)	0.94	5765.2	7.30 (0.13)	0.99
6	49.90 (0.80)	566	45.90 (8.11)	0.98	5482.8	5.10 (0.09)	0.99
Average	190.07 (3.13)	1008.13	174.52 (16.57)	0.97	5040.67	15.58 (0.34)	0.96

lower number of attributes and training instances yield the best classification rates in the test data: 1% and 0.8% respectively. On other hand, the class EC 2 has the highest error (6.1%), having the highest number of attributes. This error rate (6.1%) is considered acceptable by specialists and a quite good classification rate. Observing Table 2, columns 4 and 7, one sees that the error rate in the non class is lower than in the positive class, for all EC classes. The number of negative instances (counter examples) is higher than the positive instances (examples). This occurs because the negative instances of one class are composed by the positive instances of the others five classes.

This classification was extended to the second level of the EC hierarchy. In [14] experiments with the six subclasses of the EC 6 class were performed. All proteins from the SWISS-PROT database which are annotated with "EC 6.x" numbers were used (5640 proteins). The results regarding the global classification error for each EC 6 subclass, performing the 10-fold-cross-validation, were as follows. The average error was 1.4% and only 7.9 test instances were not correctly classified. An average accuracy (acc) of 98.6% was obtained. Regarding the class and non-class, the average confidences were approximately 0.73 and 0.99, respectively. The EC 6.6 subclass had the best classification rate, 0.2%, and the class EC 3 had the highest error (4.2%). These classification rates are better than those obtained using data from EC main classes. This is explained by the fact that data from EC 6 subclasses possesses smaller number of attributes and instances. The learning algorithms works better in such conditions, performing good generalizations on test data.

8 Multiagent Clustering

Clustering is widely used in bioinformatics, normally via classical methods. One issue with most of these methods is that they rely on central data structures. However, the current use of Internet resources in genome annotation (given distribution of data, privacy, etc.) requires new ways of dealing with data clustering. In [15] a distributed clustering algorithm, Bee clustering, is proposed, which is inspired by the organization of bee colonies. Bee clustering relies on a bee colony behavior to form groups of agents that represent data to be grouped. This behavior is known as recruitment. In nature, bees travel far away from the hive to collect nectar. They return to the hive with nectar and information about the nectar source to recruit other bees to that food source. This recruitment is performed by dancing, a behavior in which a bee communicates the directions, distance, and desirability of the food source to other bees. The proposed algorithm uses this behavior to create groups of agents. Each agent behaves like a bee and represents an object to be grouped. Hence, here an agent dances to recruit other agents to join its group. Our agents need to decide if they change groups or not.

In the bee clustering algorithm a similar, mathematical model to form groups of agents with similar features was used. Since each bee agent represents an object that needs to be grouped, the attributes of this object constitute the set of agent's features. Bees have only a limited knowledge: they only know about the agents that are placed in their groups and they cannot remember their past groups, i.e. they have no memory. Despite this, they are able to group together to form clusters in which agents with similar features are in the same group.

The main features of the algorithm are: the computation is distributed; there is no single point of failure (it does not rely on any centralized data structure); hints about the clustering (such as number of clusters, size or density of cluster) do not need to be given a priori.

Experiments were performed to investigate the quality of the proposed ant clustering algorithm using public domain data set as well as some synthetic datasets related to bioinformatics plus the dataset *Glass* from UCI (because it has also different substructures of clusterings).

The dataset used are:

- *Leukemia*: contains data on gene expression related to a subtype of leukemia. This data set is composed by 271 elements that can be divided in 2 different substructures: one with 3 classes and another with 7 classes. Each element of the set has 327 attributes.
- ds2c2sc13: contains 588 synthetic data generated so that there are 3 different substructures. These contain 2, 5 or 13 classes respectively. Each data has 2 attributes.
- *Glass*: contains 214 data, each having 10 attributes. There can be 3 substructures, having 2, 5 or 6 classes.

Table 3 shows averages and standard deviation (over 30 repetitions) regarding the *Rand* index and the number of groups that were identified by the Bee Clustering algorithm and the following other algorithms that were used as comparison: *K-means*, *average-link* and MOCLE [17]. Algorithms that were designed for multiobjective clustering (e.g. MOCLE) tend to find more than one structure in the dataset. *K-means* and *Average Link* on the other hand are not able to find diversity in those structures unless one changes their initial parameters (e. g. the k for *K-means*).

Moreover, in Table 3 one can see that Bee Clustering finds, for each dataset, one particular structure that is different from those found by the other algorithms. In the case of the *leukemia* dataset, Bee Clustering found the structure

Table 3. Average of the *Rand* index over 30 repetitions: algorithms *K-means*, *Average link*, MOCLE and *Bee Clustering*

Leukemia - 3 Classes	*K-means*	*Average Link*	MOCLE	Bee Clustering
Rand index	0.31 (0.31)	0.32 (0)	0.31 (0.02)	**0.49** (0.02)
Identified				
Number of Groups	5.6 (2)	6 (0)	7.3 (0.5)	**3.1** (0.3)
Leukemia - 7 Classes	*K-means*	*Average Link*	MOCLE	Bee Clustering
Rand index	0.75 (0.02)	0.64 (0)	**0.77** (0)	0.51 (0.09)
Identified				
Number of Groups	8.4 (1.9)	16 (0)	**8** (0)	3.1 (0.3)
ds2c2sc13 - 2 Classes	*K-means*	*Average Link*	MOCLE	Bee Clustering
Rand index	**1** (0)	**1** (0)	**1** (0)	0.52 (0.03)
Identified				
Number of Groups	**2** (0)	**2** (0)	**2** (0)	14.3 (1.7)
ds2c2sc13 - 5 Classes	*K-means*	*Average Link*	MOCLE	Bee Clustering
Rand index	0.79 (0.04)	**1** (0)	**1** (0)	0.55 (0.08)
Identified				
Number of Groups	4.7 (0.9)	**5** (0)	**5** (0)	14.3 (1.7)
ds2c2sc13 - 13 Classes	*K-means*	*Average Link*	MOCLE	Bee Clustering
Rand index	0.75 (0.02)	0.64 (0)	0.77 (0)	**0.80** (0.02)
Identified				
Number of Groups	8.4 (1.9)	16 (0)	8 (0)	**14.3** (1.7)
***Glass* - 2 Classes**	*K-means*	*Average Link*	MOCLE	Bee Clustering
Rand index	0.63 (0.02)	0.67 (0)	**0.75** (0.02)	0.57 (0.06)
Identified				
Number of Groups	3 (0.80)	9 (0)	**2** (0.70)	5.4 (0.52)
***Glass* - 5 Classes**	*K-means*	*Average Link*	MOCLE	Bee Clustering
Rand index	0.47 (0.04)	**0.56** (0)	**0.56** (0.06)	0.51 (0.02)
Identified				
Number of Groups	3 (0.80)	12 (0)	10 (3.33)	**5.4** (0.52)
***Glass* - 6 Classes**	*K-means*	*Average Link*	MOCLE	Bee Clustering
Rand index	0.23 (0.15)	0.26 (0)	0.27 (0.01)	**0.56** (0.07)
Identified				
Number of Groups	5 (2.80)	12 (0)	4 (2.80)	**5.4** (0.52)

with 3 classes (average 3.1). Besides, for this case of 3 classes in terms of *Rand* Index, Bee Clustering has outperformed the others. Similar situation happens regarding the ds2c2sc13 dataset where Bee Clustering has identified the structure with 13 classes (14.3). Bee Clustering has outperformed the others in terms of *Rand* Index for the case with 13 classes. Finally, regarding the *Glass* dataset with 6 classes, Bee Clustering has outperformed the others in terms of *Rand* index and was the only algorithm that found the 6 classes (5.4).

In summary, it was noticed that a particular feature of Bee Clustering, namely its capacity of finding the number of classes that is less obvious to the other algorithms which is a desired feature in bioinformatics where this number of classes is not known a priori (as e.g. synthetic or UCI datasets). The next step is to use a multiobjective clustering approach as proposed in [16].

9   Conclusion

DM techniques are being employed in bioinformatics with increasing success. However, some problems are still prohibitive for symbolic machine learning methods. Especially in annotation, using all available data regarding motifs as attributes has a high cost. Thus agent-based paradigms may help in the sense that the data can be divided among these agents. When conflicting rules are found, agents may negotiate and/or form an integrated model. This paper has discussed five scenarios where DM and MAS were used to produce results that facilitate the task of annotation. Other application areas are possible. In particular, ML techniques can be employed for gene recognition looking for particular signals in DNA sequences, and for gene expression analysis, where the expression level of a subset of genes has been used for tissues classification.

Acknowledgments

The author is partially supported by the Brazilian National Council for Scientific and Technological Development (CNPq). I also like to acknowledge the co-authors of related works: André Carvalho, Cássia T. dos Santos, Daniela Scherer dos Santos, and Luciana Schroeder.

References

1. Baldi, P., Søren, B.: Bioinformatics: the machine learning approach, 351 p. MIT Press, Cambridge (1998)
2. Batista, G.E.A.P.A., Monard, M.C., Bazzan, A.L.C.: Improving rule induction precision for automated annotation by balancing skewed data sets. In: López, J.A., Benfenati, E., Dubitzky, W. (eds.) KELSI 2004. LNCS, vol. 3303, pp. 20–32. Springer, Heidelberg (2004)
3. Bazzan, A.L.C., Ceroni da Silva, S., Engel, P.M., Schroeder, L.F.: Automatic annotation of keywords for proteins related to mycoplasmataceae using machine learning techniques. Bioinformatics 18(S2), S1–S9 (2002)
4. Bazzan, A.L.C., Duarte, R., Pitinga, A.N., Schroeder, L.F., Silva, S.C., Souto, F.A.: ATUCG-an agent-based environment for automatic annotation of genomes. International Journal of Cooperative Information Systems 12(2), 241–273 (2003)
5. Bazzan, A.L.C., Santos, C.T.: Selection of data sets of motifs as attributes in the process of automating the annotation of proteins' keywords in bioinformatics. In: Setubal, J.C., Verjovski-Almeida, S. (eds.) BSB 2005. LNCS (LNBI), vol. 3594, pp. 230–233. Springer, Heidelberg (2005)
6. BenHur, A., Brutlag, D.: Sequence motifs: highly predictive features of protein function. In: Feature extraction, foundations and applications. Springer, Heidelberg (2005)
7. Bryson, K., Luck, M., Joy, M., Jones, D.: Applying agents to bioinformatics in GeneWeaver. In: Klusch, M., Kerschberg, L. (eds.) CIA 2000. LNCS, vol. 1860, pp. 60–71. Springer, Heidelberg (2000)
8. Cai, C.Z., Han, L.Y., Ji, Z.L., Chen, Y.Z.: Enzyme family classification by support vector machines. Proteins: Structure, Function, and Bioinformatics 55(1), 66–76 (2004)

9. Cao, L., Luo, C., Zhang, C.: Agent-mining interaction: An emerging area. In: Gorodetsky, V., Zhang, C., Skormin, V.A., Cao, L. (eds.) AIS-ADM 2007. LNCS, vol. 4476, pp. 60–73. Springer, Heidelberg (2007)
10. Craven, M.W., Shavlik, J.W.: Machine learning approaches to gene recognition. IEEE Expert 9(2), 2–10 (1994)
11. Decker, K., Zheng, X., Schmidt, C.: A multi-agent system for automated genomic annotation. In: Proc. of the Int. Conf. Autonomous Agents, Montreal. ACM Press, New York (2001)
12. des Jardins, M., Karp, P.D., Krummenacker, M., Lee, T.J., Ouzounis, C.A.: Prediction of enzyme classification from protein sequence without the use of sequence similarity. In: Proceedings of the International Conference on Intelligent Systems Molecular Biology, pp. 92–99 (1997)
13. dos Santos, C.T., Bazzan, A.L.C.: Integrating knowledge through cooperative negotiation – a case study in bioinformatics. In: Gorodetsky, V., Liu, J., Skormin, V.A. (eds.) AIS-ADM 2005. LNCS (LNAI), vol. 3505, pp. 277–288. Springer, Heidelberg (2005)
14. dos Santos, C.T., Bazzan, A.L.C., Lemke, N.: Automatic classification of enzyme family in protein annotation. In: Proc. of the Brazilian Symposium on Bioinformatics 2009. Springer, Heidelberg (2009)
15. dos Santos, D.S., Bazzan, A.L.C.: A bee inspired clustering algorithm. In: Proc. of the 2009 IEEE Swarm Intelligence Symposium, Nashville, pp. 160–167. IEEE, Los Alamitos (2009)
16. dos Santos, D.S., de Oliveira, D., Bazzan, A.L.C.: A multiagent, multiobjective clustering algorithm. In: Cao, L. (ed.) Data Mining and Multiagent Integration. Springer, Heidelberg (2009)
17. Faceli, K., Carvalho, A.C.P.L.F., Souto, M.C.P.: Multi-objective clustering ensemble. In: Proceedings of the Sixth International Conference on Hybrid Intelligent Systems (HIS 2006), Washington, DC, USA, p. 51. IEEE Computer Society, Los Alamitos (2006)
18. Kretschmann, E., Fleischmann, W., Apweiler, R.: Automatic rule generation for protein annotation with the C4.5 data mining algorithm applied on SWISS-PROT. Bioinformatics 17, 920–926 (2001)
19. Schroeder, L.F., Bazzan, A.L.C.: A multi-agent system to facilitate knowledge discovery: an application to bioinformatics. In: Proceedings of the Workshop on Bioinformatics and Multi-Agent Systems (BIXMAS 2002), Bologna, Italy, pp. 44–50 (2002)
20. Setúbal, J.C., Meidanis, J.: Introduction to Computational Molecular Biology. PWS Publishing Company (1997)
21. Weinert, W.R., Lopes, H.S.: Neural networks for protein classification. Applied Bioinformatics 3(1), 41–48 (2004)

Knowledge-Based Reinforcement Learning for Data Mining

Daniel Kudenko and Marek Grzes

Department of Computer Science, University of York, York YO105DD, UK
{kudenko,grzes}@cs.york.ac.uk

1 Extended Abstract

Data Mining is the process of extracting patterns from data. Two general avenues of research in the intersecting areas of agents and data mining can be distinguished. The first approach is concerned with mining an agent's observation data in order to extract patterns, categorize environment states, and/or make predictions of future states. In this setting, data is normally available as a batch, and the agent's actions and goals are often independent of the data mining task. The data collection is mainly considered as a side effect of the agent's activities. Machine learning techniques applied in such situations fall into the class of supervised learning. In contrast, the second scenario occurs where an agent is actively performing the data mining, and is responsible for the data collection itself. For example, a mobile network agent is acquiring and processing data (where the acquisition may incur a certain cost), or a mobile sensor agent is moving in a (perhaps hostile) environment, collecting and processing sensor readings. In these settings, the tasks of the agent and the data mining are highly intertwined and interdependent (or even identical). Supervised learning is not a suitable technique for these cases. Reinforcement Learning (RL) enables an agent to learn from experience (in form of reward and punishment for explorative actions) and adapt to new situations, without a teacher. RL is an ideal learning technique for these data mining scenarios, because it fits the agent paradigm of continuous sensing and acting, and the RL agent is able to learn to make decisions on the sampling of the environment which provides the data. Nevertheless, RL still suffers from scalability problems, which have prevented its successful use in many complex real-world domains. The more complex the tasks, the longer it takes a reinforcement learning algorithm to converge to a good solution. For many real-world tasks, human expert knowledge is available. For example, human experts have developed heuristics that help them in planning and scheduling resources in their work place. However, this domain knowledge is often rough and incomplete. When the domain knowledge is used directly by an automated expert system, the solutions are often sub-optimal, due to the incompleteness of the knowledge, the uncertainty of environments, and the possibility to encounter unexpected situations. RL, on the other hand, can overcome the weaknesses of the heuristic domain knowledge and produce optimal solutions. In the talk we propose two techniques, which represent first steps in

L. Cao et al. (Eds.): ADMI 2009, LNCS 5680, pp. 21–22, 2009.
© Springer-Verlag Berlin Heidelberg 2009

the area of knowledge-based RL (KBRL). The first technique [1] uses high-level STRIPS operator knowledge in reward shaping to focus the search for the optimal policy. Empirical results show that the plan-based reward shaping approach outperforms other RL techniques, including alternative manual and MDP-based reward shaping when it is used in its basic form. We showed that MDP-based reward shaping may fail and successful experiments with STRIPS-based shaping suggest modifications which can overcome encountered problems. The STRIPS-based method we propose allows expressing the same domain knowledge in a different way and the domain expert can choose whether to define an MDP or STRIPS planning task. We also evaluated the robustness of the proposed STRIPS-based technique to errors in the plan knowledge. In case that STRIPS knowledge is not available, we propose a second technique [2] that shapes the reward with hierarchical tile coding. Where the Q-function is represented with low-level tile coding, a V-function with coarser tile coding can be learned in parallel and used to approximate the potential for ground states. In the context of data mining, our KBRL approaches can also be used for any data collection task where the acquisition of data may incur considerable cost. In addition, observing the data collection agent in specific scenarios may lead to new insights into optimal data collection behaviour in the respective domains. In future work, we intend to demonstrate and evaluate our techniques on concrete real-world data mining applications.

References

1. Grzes, M., Kudenko, D.: Plan-based Reward Shaping for Reinforcement Learning. In: Fourth International IEEE Conference on Intelligent Systems, vol. 2, pp. 22–29 (2008)
2. Grzes, M., Kudenko, D.: Learning potential for reward shaping in reinforcement learning with tile coding. In: Proceedings AAMAS 2008 Workshop on Adaptive and Learning Agents and Multi-Agent Systems (ALAMAS-ALAg 2008), pp. 17–23 (2008)

Ubiquitous Intelligence in Agent Mining

Longbing Cao, Dan Luo, and Chengqi Zhang

Faculty of Engineering and Information Technology,
University of Technology Sydney, Australia
{lbcao,dluo}@it.uts.edu.au

Abstract. Agent mining, namely the interaction and integration of multi-agent and data mining, has emerged as a very promising research area. While many mutual issues exist in both multi-agent and data mining areas, most of them can be described in terms of or related to ubiquitous intelligence. It is certainly very important to define, specify, represent, analyze and utilize ubiquitous intelligence in agents, data mining, and agent mining. This paper presents a novel but preliminary investigation of ubiquitous intelligence in these areas. We specify five types of ubiquitous intelligence: data intelligence, human intelligence, domain intelligence, network and web intelligence, organizational intelligence, and social intelligence. We define and illustrate them, and discuss techniques for involving them into agents, data mining, and agent mining for complex problem-solving. Further investigation on involving and synthesizing ubiquitous intelligence into agents, data mining, and agent mining will lead to a disciplinary upgrade from methodological, technical and practical perspectives.

1 Introduction

There is an increasingly evident need of integrating agents (namely multi-agent systems) and data mining (knowledge discovery from data) for complex problem-solving in both agents and data mining areas. Agent mining [1,2,3,6,11,16,15,19], namely agents and data mining interaction and integration[1], has emerged to be a new and promising discipline. Great efforts have been made on agent mining from aspects of theoretical foundation-building, technological fundamentals, and technical means and tools. More and more applications have been reported benefiting from the synergy of agents and data mining.

In agent mining, a critical issue is to deal with those issues commonly seen in both agents and data mining areas. Here are some examples. Both agents and data mining involve aspects such as domain knowledge, constraints, human roles and interaction, lifecycle and process management, organizational and social factors. Many agent and data mining systems are dynamic and need to cater for online, run-time and ad-hoc requests. With the involvement of social intelligence and complexities, both areas care about reliability, reputation, risk, privacy, security, trust, and outcome actionability. Research in one area can actually stimulate, complement and enhance that in the other.

---

[1] AMII-SIG: www.agentmining.org

L. Cao et al. (Eds.): ADMI 2009, LNCS 5680, pp. 23–35, 2009.

A typical understanding of the above mutual issues existing in agents and data mining is from the angle of ubiquitous intelligence. Ubiquitous intelligence surrounds a real-world agent mining problem can be identified and categorized into the following types.

- Data intelligence,
- Human intelligence,
- Domain intelligence,
- Network intelligence,
- Organizational intelligence, and
- Social intelligence.

Furthermore, for agents, data mining and agent mining, it is necessary to not only involve an individual type of the above intelligence, but also consolidate the relevant ubiquitous intelligence into the modeling, evaluation, working process and systems. In this paper, we discuss the concepts and aims of involving each of the intelligence, and the corresponding techniques and case studies for involving them into agents, data mining, and agent mining.

The listed ubiquitous intelligence and the consolidation in a system open a new angle of observing key and mutual challenges in agents, data mining, and agent mining. Following this direction, we believe many issues in the above areas can be addressed or with solutions provided. As a result, many great opportunities will emerge with more advanced, effective and efficient methodologies, techniques, means and tools and applications in dealing with complex multi-agent, data mining and agent mining problems and systems.

2   Data Intelligence

2.1   What Is Data Intelligence

Definition 1. *(Data Intelligence) tells interesting stories and/or indicators hidden in data about a business problem. The intelligence of data emerges in the form of interesting patterns and actionable knowledge.*

There are two levels of data intelligence:

- *General level of data intelligence*: refers to the knowledge identified from explicit data, presenting general knowledge about a business problem, and
- *In-depth level of data intelligence*: refers to the knowledge identified in more complex data, using more advanced techniques, or disclosing much deeper information and knowledge about a problem.

Taking association rule mining as an example, a general level of data intelligence is frequent patterns identified in basket transactions, while *associative classifiers* reflect deeper level of data intelligence.

2.2   Aims of Involving Data Intelligence

We aim to disclose data intelligence from multiple perspectives. One of the angles to observe data intelligence is the data explicitness or implicitness.

- *Explicit data intelligence*, refers to the level of data intelligence disclosing explicit characteristics or exhibited explicitly. An example of explicit data intelligence is the trend of a stock market index or of a stock price dynamics.
- *Implicit data intelligence*, refers to the level of data intelligence disclosing implicit characteristics or exhibited implicitly. In stock markets, an example of implicit data intelligence is the trading behavior patterns of a hidden group in which investors are associated with each other.

Both explicit data intelligence and implicit data intelligence may present intelligence at either general or in-depth level.

Another angle of scrutinizing data intelligence is from either a syntactic or a semantic perspective.

- *Syntactic data intelligence*, refers to the kind of data intelligence disclosing syntactic characteristics. An example of syntactic data intelligence is itemset associations.
- *Semantic data intelligence*, refers to the kind of data intelligence disclosing semantic characteristics. An example of semantic data intelligence is temporal trading behavior embedding temporal logic relationship amongst trading behaviors.

Similarly, both syntactic data intelligence and semantic data intelligence may present intelligence at either general or in-depth level.

2.3   Aspects of Data Intelligence

Even though the mainstream agents and data mining focus on substantial investigation of varying data for hidden interesting patterns or knowledge, the real-world data and its surroundings are usually much more complicated. The following lists aspects that may be associated with data intelligence.

- Data type such as numeric, categorical, XML, multimedia and composite data
- Data timing such as temporal and sequential
- Data spacing such as spatial and temporal-spatial
- Data speed and mobility such as high frequency, high density, dynamic data and mobile data
- Data dimension such as multi-dimensional, high-dimensional data, and multiple sequences
- Data relation such as multi-relational, linkage record
- Data quality such as missing data, noise, uncertainty, and incompleteness
- Data sensitivity like mixing with sensitive information

Deeper and wider analysis is required to disclose in-depth data intelligence in complex data. Two kinds of efforts: data engineering and data mining, need to be further developed for processing and analyzing real-world data complexities such as multi-dimensional data, high-dimensional data, mixed data, distributed data, and processing and mining unbalanced, noisy, uncertain, incomplete, dynamic, and stream data.

3 Domain Intelligence

3.1 What Is Domain Intelligence

Definition 2. *(Domain Intelligence) refers to the intelligence that emerges from the involvement of domain factors and resources into agents, data mining and agent mining, which wrap not only a problem but its target data and environment. The intelligence of domain is embodied through the involvement into modeling process, models and systems.*

Domain intelligence involves qualitative and quantitative aspects. They are instantiated in terms of aspects such as domain knowledge, background information, prior knowledge, expert knowledge, constraints, organizational factors, business process, workflow, as well as environmental aspects, business expectation and interestingness.

3.2 Aims of Involving Domain Intelligence

Multiple types of domain intelligence may be engaged in agents, data mining and agent mining.

- *Qualitative domain intelligence*, refers to the type of domain intelligence that discloses qualitative characteristics or involves qualitative aspects. Taking stock data mining as an example, fund managers have qualitative domain intelligence such as "beating the market", when they evaluate the value of a trading pattern.
- *Quantitative domain intelligence*, refers to the type of domain intelligence that discloses quantitative characteristics or involves quantitative aspects. An example of quantitative domain intelligence in stock data mining is whether a trading pattern can "beat VWAP[2]" or not.

The roles of involving domain intelligence in agents, data mining and agent mining are multi-form.

- Assisting in the modeling and evaluation of the problem. An example is "my trading pattern can beat the market index return" when domain intelligence of "beat market index return" is applied to evaluate a trading pattern.
- Making mining realistic and business-friendly. By considering domain knowledge, we are able to work on an actual business problem rather than an artificial one abstracted from an actual problem.

3.3 Aspects of Domain Intelligence

In the mainstream of agents and data mining, the consideration of domain intelligence is mainly embodied through involving domain knowledge, prior knowledge, or mining the process and/or workflow associated with a business problem.

In a specific domain problem, domain intelligence may be presented in multiple aspects, for instance, some of the following aspects:

---

[2] VWAP is a trading acronym for Volume-Weighted Average Price, the ratio of the value traded to total volume traded over a particular time horizon.

- Domain knowledge,
- Background and prior information,
- Meta-knowledge and meta-data
- Constraints,
- Business process,
- Workflow,
- Benchmarking and criteria definition, and
- Business expectation and interest.

4 Network Intelligence

4.1 What Is Network Intelligence

Definition 3. *(Network Intelligence) refers to the intelligence that emerges from both web and broad-based network information, facilities, services and processing surrounding an agent, data mining or agent mining problem and system.*

Network intelligence involves both *web intelligence* and *broad-based network intelligence* such as information and resources distribution, linkages amongst distributed objects, hidden communities and groups, web service techniques, messaging techniques, mobile and personal assistant agents for decision-support, information and resources from network, and in particular the web, information retrieval, searching and structuralization from distributed and textual data. The information and facilities from the networks surrounding the target business problem either consist of the problem constituents, or can contribute to useful information for actionable knowledge discovery and delivery in agents, data mining and agent mining.

4.2 Aims of Involving Network Intelligence

The aims of involving network intelligence into agents, data mining and agent mining include multiple aspects, for example, to

- Involve data and information from a community or team
- Involve and mine web data,
- Involve and mine network data,
- Support pattern mining and decision-making,
- Support decision-making on top of mined patterns and knowledge, and
- Support social agent and data mining systems by providing facilities for social interaction in a team.

In particular, we care about

- Discovering the business intelligence in networked data related to a business problem, for instance, discovering market manipulation patterns in cross-markets.
- Discovering networks and communities existing in a business problem and its data, for instance, discovering hidden communities in a market investor population.
- Involving networked constituent information in pattern mining on target data, for example, mining blog opinion for verifying market abnormal trading.
- Utilizing networking facilities to pursue information and tools for actionable knowledge discovery, for example, involving mobile agents to support distributed and peer-to-peer mining.

4.3 Aspects of Network Intelligence

In saying network intelligence, on one hand, we expect to fulfill the power of web and network information and facilities for agents, data mining and agent mining in terms of many aspects, for instance,

- Information and resource distribution
- Linkages amongst distributed objects
- Hidden communities and groups
- Information and resource from network and in particular the web
- Information retrieval
- Structuralization and abstraction from distributed textual (blog) data
- Distributed computing
- Web network communication techniques
- Web-based decision-support techniques
- Dynamics of networks and the web
- Multiagent-based messaging and mining

On the other hand, we focus on disclosing web and network intelligence. In this regard, there are many emergent topics to be studied. We list a few here.

- Social network mining
- Hidden group and community mining
- Context-based web mining
- Opinion formation and evolution dynamics
- Distributed and multiple source mining
- Mining changes and dynamics of network
- Multiagent-based distribute data mining

5 Human Intelligence

5.1 What Is Human Intelligence

Definition 4. *(Human Intelligence) refers to (1) explicit or direct involvement of human knowledge or human as a problem-solving constituent, etc., and (2) implicit or indirect involvement of human knowledge or human as a system component.*

Explicit or direct involvement of human intelligence may consist of human empirical knowledge, belief, intention, expectation, run-time supervision, evaluation, and an individual end user or expert groups. An example of explicit human intelligence is for a domain expert to tune parameters via user interfaces. On the other hand, implicit or indirect involvement of human intelligence may present as imaginary thinking, emotional intelligence, inspiration, brainstorm, reasoning inputs, and embodied cognition like convergent thinking through interaction with other members in assessing identified patterns. Examples of involving implicit human intelligence are user modeling for game behavior design, collecting opinion from an expert group for guiding model optimization, and utilizing embodied cognition for adaptive model adjustment.

5.2   Aims of Involving Human Intelligence

The importance of involving human into data mining has been widely recognized. With the systematic specification of human intelligence, we are able to convert agent mining toward more human-centered, interactive, dynamic and user-friendly, enhancing the capability of dealing with complex agent mining issues, forming closed-loop agent mining systems, and strengthening the usability of agent mining.

- Human-centered capability: The involvement of human, including individual and group knowledge, experience, preferences, cognition, thinking, reasoning etc. and more broad aspects linking to social and cultural factors (we will further expand this in social intelligence), makes it possible for utilizing human intelligence into enhancing agent mining capability. Based on the depth and breadth of human involvement, the cooperation of human with agent mining can be human-centered or human-assisted;
- Interactive capability: The involvement of human is through interactive interfaces. This forms interactive agent mining capability and systems, to effectively and sufficiently cater for human intelligence into agent mining;
- Improving adaptive capability: Real-life agent mining applications are often dynamic. Agent mining models are often pre-defined and cannot adapt to the dynamics. The involvement of human intelligence can assist with the understanding and capture of such dynamics and change, and guide the corresponding adjustment and retraining of models;
- User-friendliness: The catering of user preferences, characteristics, and requests in agent mining will certainly make it more user-friendly;
- Dealing with complex agent mining issues: Many complex issues cannot be handled very well without the involvement domain experts. Complex knowledge discovery from complex data can benefit from inheriting and learning expert knowledge, enhancing the understanding of domain, organizational and social factors through expert guidelines, embedding domain experts into agent mining systems, etc.
- Closed-loop agent mining: In general, agent mining systems are open. As we learn from disciplines such as cybernetics, problem-solving systems are likely to be closed-loop in order to deal with environmental complexities and to achieve robust and dependable performance. This is the same for actionable knowledge discovery and delivery systems. The involvement of human can essentially contribute to closed-loop agent mining.
- Enhancing usability of agent mining: Driven by the involvement of human intelligence and the corresponding development and support, the usability of agent mining systems can be greatly enhanced. Usability measures the quality of a user's experience when interacting with an agent mining system.

5.3   Aspects of Human Intelligence

The aspects of human intelligence in agents, data mining and agent mining are embodied through many ways.

- Human empirical knowledge,
- Belief, intention, expectation,
- Sentiment, opinion
- Run-time supervision, evaluation,
- Expert groups
- Imaginary thinking,
- Emotional intelligence,
- Inspiration,
- Brainstorm,
- Retrospection,
- Reasoning inputs, and
- Embodied cognition like convergent thinking through interaction with other members in assessing identified patterns

6   Organizational Intelligence

6.1   What Is Organizational Intelligence

Definition 5. *(Organizational Intelligence) refers to the intelligence that emerges from involving organization-oriented factors and resources into agents, data mining and agent mining. The organizational intelligence is embodied through its involvement into the system process, modeling and systems.*

6.2   Aims of Involving Organizational Intelligence

In a complex organization, the involvement of organizational intelligence is essential in many aspects, for instance,

- Reflecting the organization reality, needs and constraints in business modeling and finding delivery
- Satisfying organizational goals and norms, policies etc. regulation and convention,
- Considering the impact of organizational interaction and dynamics in the modeling and deliverable design,
- Catering for organizational structure and its evolution in data extraction, preparation, modeling, and delivery.

6.3   Aspects of Organizational Intelligence

Organizational intelligence consist of many aspects, for example

- Organizational structures related to key issues such as where data comes from and who in which branch needs the findings
- Organizational behavior related to key issues such as the business and data understanding and finding delivery of how individuals and groups act in an organization
- Organizational evolution and dynamics related to key issues such as data and information change, affecting model/pattern/knowledge evolution and adaptability

- Organizational/business regulation and convention related to key issues such as business understanding and finding delivery, including rules, policies, protocols, norms, law etc.
- Business process and workflow related to key issues such as data (reflecting process and workflow) and business understanding, goal and task definition, and finding delivery etc.
- Organizational goals related to key issues such as problem definition, goal and task definition, performance evaluation, etc.
- Organizational actors and roles related to key issues such as system actor definition, user preferences, knowledge involvement, interaction, interface and service design, delivery etc.
- Organizational interaction related to key issues such as data and information interaction amongst sub-systems and components, data sensitivity and privacy, and interaction rules applied on organizational interaction that may affect data extraction, integration and processing, and pattern delivery and so on.

7   Social Intelligence

7.1   What Is Social Intelligence

Definition 6. *(Social Intelligence) refers to the intelligence that emerges from the group interactions, behaviors and corresponding regulation surrounding an agent mining problem.*

Social intelligence covers both *human social intelligence* and *animat/agent-based social intelligence*. *Human social intelligence* is related to aspects such as social cognition, emotional intelligence, consensus construction, and group decision. *Animat/agent-based social intelligence* involves swarm intelligence, action selection and the foraging procedure. Both sides also engage social network intelligence, collective interaction, as well as social regulation rules, law, trust and reputation for governing the emergence and use of social intelligence.

7.2   Aims of Involving Social Intelligence

In designing complex problem-solving systems in social environment, both human social intelligence and agent-based social intelligence may play an important role, for instance,

- Enhancing social computing capability of agent mining methods and systems,
- Implementing agent mining and evaluation in a social and group-based manner, under supervised or semi-supervised condition,
- Utilizing social group thinking and intelligence emergence into complex agent mining problem-solving,
- Building social agent mining software on the basis of software agents, to facilitate human-mining interaction, group decision-making, self-organization and autonomous action selection by data mining agents. This may benefit from multi-agent data mining and warehousing,

- Defining and evaluating social performance including trust and reputation in developing quality social agent mining software, and
- Enhancing agent mining project management and decision-support capabilities of the identified findings in a social environment.

7.3   Aspects of Social Intelligence

Aspects of social intelligence are in multiple forms. We illustrate them from human social intelligence and animat/agent-based social intelligence respectively.

Human social intelligence aspects consist of aspects such as *social cognition, emotional intelligence, consensus construction*, and *group decision*.

- *Social cognition*, aspects related to how a group of people process and use social information, which can inform how to involve what information into agent mining,
- *Emotional intelligence*, aspects related to a group of people's emotions and feelings, which can inform interface and interaction design, performance evaluation and finding delivery for agent mining.
- *Consensus construction*, aspects related to how a group of people think and how thinking evolves in a group toward a convergence, in particular, under a divergent thinking situation, which can inform the conflict resolution if people with different background value different aspects in pattern selection or if there is a conflict between technical and business interest,
- *Group decision*, aspects related to strategies and methods used by a group of people in making a decision, which can inform the discussion between business modelers and end users.

Animat/agent-based social intelligence aspects consist of

- Swarm/collective intelligence, aspects related to collaboration and competition of a group of agents in handling a social data mining problem, which can assist in complex data mining through multi-agent interaction, collaboration, coordination, negotiation and competition, and
- Behavior/group dynamics, aspects related to group formation, change and evolution, and group behavior dynamics, which can assist in simulating and understanding structure, behavior and impact of mining a group/community.

In addition, both human/agent social intelligence also involves many common aspects, such as

- Social network intelligence,
- Collective interaction,
- Social behavior network,
- Social interaction rules, protocols, norms etc.,
- Trust and reputation etc. and
- Privacy, risk, security etc. in social context.

8   Discussions

The involvement of ubiquitous intelligence is very important for handling open complex problems, such as open complex intelligent systems [8] and open complex agent systems [9]. Besides the individual engagement of each type of intelligence, more critical problem is to synthesize them into a problem-solving system. This comes to an interesting but difficult problem, namely how they can be integrated. Methodologies and approaches for doing so are not currently mature, just as the studies on the five proposed types of intelligence.

We have had a few trials in this regard, which are essentially helpful for detailing ubiquitous intelligence in constructing open complex agent systems, data mining systems, and agent mining systems.

- Intelligence meta-synthesis [17,18,14]: This is a methodology proposed to deal with open complex giant systems [17,18]. It proposes a general framework of integrating human intelligence, computer intelligence into problem-solving systems, in particular, by establishing a hall for workshop of meta-synthesis from qualitative to quantitative. Agent-based prototypes have been built for applying metasynthesis in handling macro-economic decision-making [13].
- Metasynthetic computing [9]: This is a computing technique proposed to construct systems, mainly open complex intelligent systems, by utilizing the methodology of intelligence meta-synthesis. A typical solution is to integrate agents, services, organizational and social computing for handling open complex systems [10], by establishing an *m-space* powered by *m-computing* and *m-interaction* [5].

A typical example of involving ubiquitous intelligence into data mining is *domain driven data mining* [4] for actionable knowledge discovery and delivery [12].

There are many open issues in involving and integrating ubiquitous intelligence into agents, data mining, and agent mining. We believe that the studies on this will lead to great opportunities for innovative methodologies, techniques, means and tools, as well as applications in the relevant areas including agents, data mining, and agent mining. This will consequently and definitely promote the transfer of the relevant disciplines toward more advanced, effective and efficient stage from methodological, technical and practical aspects.

9   Conclusion

Agents, data mining and agent mining face many critical mutual issues. In this paper, we propose the concept of ubiquitous intelligence to describe and address these issues. They are

- Data intelligence, refers to both general level and in-depth level of data intelligence from both syntactic and semantic perspectives;
- Human intelligence, refers to both explicit or direct and implicit or indirect involvement of human intelligence;
- Domain intelligence, refers to both qualitative and quantitative domain intelligence;

- Network intelligence, refers to both web intelligence and broad-based network intelligence,
- Organizational intelligence, refers to organizational goals, structures, rules, and dynamics, and
- Social intelligence, refers to both human social intelligence and agent/animat-based social intelligence.

We have discussed their definition, aims, aspects and techniques for involving them into agents, data mining, and agent mining.

In our other works, preliminary examples and case studies have been presented to illustrate these concepts. In particular, we have proposed a methodology, namely *intelligence meta-synthesis* and *metasynthetic computing* for synthesizing ubiquitous intelligence into domain-driven actionable knowledge delivery.

Acknowledgements

This work is sponsored in part by Australian Research Council Discovery Grants (DP0988016, DP0773412, DP0667060) and ARC Linkage Grant (LP0989721, LP0775041).

References

1. Cao, L., Gorodetsky, V., Mitkas, P.: Agent Mining: The Synergy of Agents and Data Mining. IEEE Intelligent Systems (May/June 2009)
2. Cao, L., Gorodetsky, V., Mitkas, P.: Guest Editors' Introduction: Agents and Data Mining (May/June 2009)
3. Cao, L. (ed.): Data Mining and Multiagent Integration. Springer, Heidelberg (2009)
4. Cao, L., Yu, P., Zhang, C., Zhao, Y.: Domain Driven Data Mining. Springer, Heidelberg (2009)
5. Cao, L., Dai, R., Zhou, M.: Metasynthesis: M-Space, M-Interaction and M-Computing for Open Complex Giant Systems. IEEE Trans. on Systems, Man, and Cybernetics–Part A (2009)
6. Cao, L., et al.: Editor's Introduction: Interaction between Agents and Data Mining. Int'l. J. Intelligent Information and Database Systems 2(1), 15 (2008)
7. Cao, L., Zhang, H., Zhao, Y., Zhang, C.: General Frameworks for Combined Mining: Case Studies in e-Government Services. Submitted to ACM TKDD (2008)
8. Cao, L., Dai, R.: Open Complex Intelligent Systems. Post & Telecom Press (2008)
9. Cao, L.: Metasynthetic Computing for Solving Open Complex Problems. In: IEEE International Workshop on Engineering Open Complex Systems: Metasynthesis of Computing Paradigms, joint with COMPSAC 2008 (2008)
10. Cao, L.: Integrating agents, services, organizational and social computing. Int. J. on Software Engineering and Knowledge Engineering (2008)
11. Cao, L., Luo, C., Zhang, C.: Agent-Mining Interaction: An Emerging Area. In: Gorodetsky, V., Zhang, C., Skormin, V.A., Cao, L. (eds.) AIS-ADM 2007. LNCS (LNAI), vol. 4476, pp. 60–73. Springer, Heidelberg (2007)
12. Cao, L., et al.: Domain-Driven actionable knowledge discovery. IEEE Intelligent Systems 22(4), 78–89 (2007)

13. Cao, L., Dai, R.: Agent-Oriented Metasynthetic Engineering for Decision Making. International Journal of Information Technology and Decision Making 2(2), 197–215 (2003)
14. Dai, R.: Qualitative-to-Quantitative Metasynthetic Engineering. Pattern Recognition and Artificial Intelligence 6(2), 60–65 (1993)
15. Gorodetsky, V., et al. (eds.): AIS-ADM 2005. LNCS (LNAI), vol. 3505. Springer, Heidelberg (2005)
16. Gorodetsky, V., et al. (eds.): AIS-ADM 2007. LNCS (LNAI), vol. 4476. Springer, Heidelberg (2007)
17. Qian, X., Yu, J., Dai, R.: A New Scientific Field–Open Complex Giant Systems and the Methodology. Chinese Journal of Nature 13(1), 3–10 (1990)
18. Qian, X.: Revisiting Issues on Open Complex Giant Systems. Pattern Recognition and Artificial Intelligence 4(1), 5–8 (1991)
19. Symeonidis, A., Mitkas, P.: Agent Intelligence through Data Mining. Springer, Heidelberg (2005)

Agents Based Data Mining and Decision Support System

Serge Parshutin and Arkady Borisov

Riga Technical University, Institute of Information Technology, 1 Kalku Str., Riga, Latvia, LV-1658
serge.parshutin@rtu.lv, arkadijs.borisovs@cs.rtu.lv

Abstract. Production planning is the main aspect for a manufacturer affecting an income of a company. Correct production planning policy, chosen for the right product at the right moment in the product life cycle (PLC), lessens production, storing and other related costs. This arises such problems to be solved as defining the present a PLC phase of a product as also determining a transition point - a moment of time (period), when the PLC phase is changed.

The paper presents the Agents Based Data Mining and Decision Support system, meant for supporting a production manager in his/her production planning decisions. The developed system is based on the analysis of historical demand for products and on the information about transitions between phases in life cycles of those products. The architecture of the developed system is presented as also an analysis of testing on the real-world data results is given.

Keywords: Software Agents, Data Mining, Decision Support, Forecasting Transition Points.

1 Introduction

Constantly evolving computer technologies are more and more becoming an inherent part of successful enterprises management and keeping its activity at a high level. Different institutions are trying to reduce their costs by fully automatising certain stages of manufacturing process as well as introducing various techniques intended for that impact general manufacturing process. Different statistical methods are employed as well, though an increasing interest in computational intelligence technologies and their practical application can be observed ever more.

We are focusing our research on studying the ability to create the Software Agents Based system that autonomously or semi-supervised will apply Data Mining and Decision Support technologies to a real-world problem. A software agent definitions and features can be found in sources [4,11]. Application examples of software agents can be found in various fields of science, industry, environment monitoring and others [1,15].

A task of product life cycle phase transition point forecasting can serve as an example of such problem where both Data Mining and Decision Support technologies should be applied. From the viewpoint of the management it is important to know, in which particular phase the product is. One of applications of that knowledge is selection of the production planning policy for the particular phase [12]. For example, for the maturity phase in case of determined demand changing boundaries it is possible to apply cyclic

L. Cao et al. (Eds.): ADMI 2009, LNCS 5680, pp. 36–49, 2009.

planning [2], whereas for the introduction and decline phase an individual planning is usually employed. As the technologies are evolving, the variability of products grows, making manual monitoring of PLCs a difficult and costly task for companies. Thus having an autonomous Agents Based system that monitors market data and creates and automatically updates lists of products for what it is reasonable to consider a production planning policy update or replacement, is one valuable alternative. This paper proposes a model of Agents Based system that ensures the solving of the aforementioned task as well as provides an analysis of system testing results.

2   Problem Statement

Any created product has a certain life cycle. The term "life cycle" is used to describe a period of product life from its introduction on the market to its withdrawal from the market. Life cycle can be described by different phases: traditional division assumes such phases like introduction, growth, maturity and decline [10]. For products with conditionally long life cycle, it is possible to make some simplification, merging introduction and growth phases into one phase - introduction.

An assumption that three different phases, namely, introduction, maturity and end-of-life are possible in the product life cycle, gives us two possible transitions . The first transition is between introduction and maturity phases and the second - between maturity and product's end-of-life.

From the side of data mining [3,5,6] , information about the demand for a particular product is a discrete time series, in which demand value is, as a rule, represented by the month. A task of forecasting a transition points between life cycle phases may be formulated as follows. Assume that $D = \{d_1,\ldots,d_i,\ldots,d_n\}$ is a dataset and $d = \{a_1,\ldots,a_j,\ldots,a_l\}$ is a discrete time series whose duration equals to l periods, where $l \in L = \{l_1,\ldots,l_h,\ldots,l_s\}$ and varies from record to record in the dataset D. For simplification, the index of d is omitted. Time series d represents a particular phase of a product life cycle, say introduction. Assume that for a particular transition, like introduction to maturity, a set of possible transition points $P = \{p_1,\ldots,p_k,\ldots,p_m\}$ is available. Having such assumptions the forecasting of a transition point for a new product, represented by a time series $d' \notin D$, will start with finding an implication between historical datasets D and P, $f : D \rightarrow P$; followed by application of found model to new data.

3   Structure of the System

The developed system contains three main elements - Data Management Agent , Data Mining Agent and Decision Analysis Agent , shown in Figure 1.

Data Management Agent. The Data Management Agent performs several tasks of managing data. It has a link to the database that contains the product demand data and regularly is updated. The Data Management Agent handles the preprocessing of the data - data normalization, exclusion of obvious outliers and data transformation to defined format before sending it to the Data Mining Agent. Such preprocessing allows

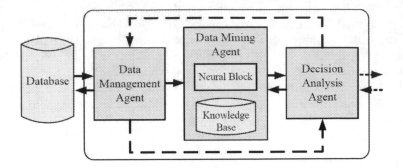

Fig. 1. Structure of the system

to lessen the impact of noisiness and dominance of data [14,16]. The transformed data record displays the demand for a product, collected within known period of time, the length of which is set by the system - day, week, month, etc. Each data record is marked with one or both markers - transition indicators ; namely, *M1* indicates the period when product switched from Introduction phase to Maturity phase; and *M2* indicates the period when product switched from Maturity phase to End-of-Life phase. Marks on transitions can guarantee that a model will be build; if we have patterns of transitions in historical data, then, theoretically, in presence of a model for generalisation, we are able to recognise those patterns in new data. Assigning markers to the demand time series is one process that currently is done manually by a human expert.

As the database is regularly updated, the Data Management Agent monitors the new data and at a defined moment of time forwards the subset of new data to the Data Mining Agent for updating the knowledge base.

Data Mining Agent. The actions, performed by the Data Mining Agent, cover such processes as initialization of a training process, performing system training and testing processes, imitation of On-line data flowing during system training and testing, Knowledge base creation and maintaining it up-to-date.

The knowledge base of the Data Mining Agent is based on the modular self-organising maps, placed in the Neural Block. Each neural net either can handle records, with equal duration l or can proceed with time series with different durations. Which option will be selected depends on the total load distribution policy, currently defined by the user. Let us illustrate the load distribution. Assume that the duration of discrete time series in dataset D varies from 10 to 16 periods, thus $l \in L = \{10, 11, \ldots, 16\}$. In this case, total system load will consist of seven values that time series duration can take. Let us assume that the load has to be distributed uniformly over modules at the condition that an individual load on separate module should not exceed four values that is $q = 4$. Under such conditions, two neural networks will be created by the moment of system initialisation. The first neural network will process time series of duration $l \in \{10, 11, 12, 13\}$; the remaining time series with duration of 14, 15 and 16 periods will be sent to the second neural net.

The option of having the On-line data flowing procedure becomes necessary due to specifics of a chosen system application environment. In the real-world situation, the information about the demand for a new product is becoming available gradually: after a regular period finishes, new demand data appear. Due to that the system must be trained to recognize transition points that will occur in the future having only several first periods of the demand time series available.

The algorithm employed is executed taking into account the following details. Pre-processed by the Data Management Agent time series d with duration l, containing demand data within introduction or maturity phase and the appropriate marker ($M1$ or $M2$ respectively) with value p, is sent to the Data Mining Agent . Having that minimal duration of a time series should be l_{min} and greater, the algorithm of On-line data flowing procedure will include these steps:

1. Define $l^* = l_{min}$;
2. Process first l^* periods of a record d with a marker value p;
3. If $l^* < l$ then increase value of l^* by one period and return to step 2; else proceed to step 4;
4. End processing of record d.

According to the chosen policy of system load distribution the fraction of a time series with marker is directed to the neural net responsible for processing the time series of specific duration.

Each self-organising map is based on the modified Kohonen map. The number of synaptic weights of a neuron in the classical Kohonen map equals to the length of records - the duration of a time series, in the input dataset [7,9,13]. Due to such limitations, it is possible to use the classical Kohonen map in the developed system only when $q = 1$ for each neural network . For to be able to maintain the system functionality while $q > 1$, it is necessary to apply some modifications to the classical Kohonen map.

A heuristic, based on substituting the distance measure, is considered in this work. We propose a substitution of a Euclidean distance measure with a measure based on Dynamic Time Warping (DTW). Using DTW allows to process time series with different duration by one neural net. Classical Dynamic Time Warping algorithm, that is applied in the system, is fully described in [8].

At the training stage, the neural networks are organised, using the data, prepared by the Data Management Agent. This is followed by forming the clusters, found by the neural networks, which will form the main part of the knowledge base.

Decision Analysis Agent. The Decision Analysis Agent is the element that uses the knowledge base of the Data Mining Agent to forecast a transition points for new products as also to analyse the alternatives of using cyclic or non-cyclic planning policies for particular products. The Decision Analysis Agent follows a certain algorithms, whose examples are provided in the next subsection.

3.1 System Functioning Algorithm

The system functioning algorithm consists of two main stages - System Learning and System Application. The System Learning stage contains two inner steps - System Training followed by System Testing and Validation. The System Application stage is the stage when the Decision Analysis Agent receives requests from the user and following certain algorithms, described in current subsection, provides the user with certain information. The system functioning algorithm may be observed in Figure 2.

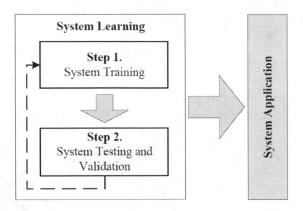

Fig. 2. System functioning algorithm

System Learning. The system learning process is fired when the Data Management Agent sends prepared data to the Data Mining Agent with "Start initial learning" or "Update" command. The "Start initial learning" command is sent when the system is launched for the first time. This will fire the process of determination and setting of basic system's parameters: the number of neural networks in the Neural Block, dimensionality and topology of neural networks and the number of synapses each neuron will have in each network. The number of networks in the Neural Block, n, is calculated empirically. Given a policy assuming uniform distribution of general load among the networks, formula (1) can be used to calculate the number of networks in the Neural Block.

$$n = \left\lceil \frac{|L|}{q} \right\rceil \ , \tag{1}$$

where q - each network's individual load; $\lceil \cdot \rceil$ - symbol of rounding up.

After the n is calculated, for each neural network m_i an interval of time series durations $[l_{i,min}; l_{i,max}]$ is set. The records with duration $l \in [l_{i,min}; l_{i,max}]$ will be processed by module m_i. Given a uniform load distribution, equation (2) can be used.

$$\begin{cases} i = 1, & l_{i,min} = l_{min} \ , \\ i > 1, & l_{i,min} = l_{i-1,max} + 1 \ ; \end{cases} \tag{2}$$

$$l_{i,max} = l_{i,min} + q - 1 \ .$$

The number of neurons of each network is determined empirically depending on the task stated. The number of synapses of a neuron in each network can be calculated using formula (3) below:

$$b_{j,i} = \left\lceil \frac{l_{i,min} + l_{i,max}}{2} \right\rceil \ , \tag{3}$$

where $b_{i,j}$ is the number of synapses of a neuron j in the network m_i.

As an alternative, a median calculation can be used for determining the number of synapses of a neuron. Assuming that network m_i can process discrete time series of duration $l \in [l_1, l_2, \ldots, l_i, \ldots, l_k]$, let us denote for each value l the number of records in the training set (a data set, forwarded by the Data Management Agent to the Data Mining Agent with "Start initial learning" command) having duration equal to l, as $f \in [f_1, f_2, \ldots, f_i, \ldots, f_k]$. By having such an assumption, a median of time series durations a network m_i can process, may be calculated with formula (4).

$$Median = \frac{\sum_{i=1}^{k} (l_i \cdot f_i)}{\sum_{i=1}^{k} f_i} \quad . \tag{4}$$

The calculated median must be rounded to the integer thus obtaining the number of synaptic connections of a neuron. Will the median be rounded to a smaller or a greater value, to a large extent depends on the task to be solved.

As the main parameters are set, the initialisation of the system begins. Synaptic weights of each neuron in each network are assigned initial values - usually small values produced by random number generator. At this moment networks in the Neural Block have no organization at all. Then the following main processes of the training step are launched: Competition, Cooperation and Synaptic adaptation.

Data Mining Agent processes each of the received records imitating the On-line data flowing. Fractions of time series are forwarded to the corresponding neural network m_i, defined by the $[l_{i,min}; l_{i,max}]$ interval, calculated with equation (2). Then the process of neuron competition for the right to become the winner (or best-matching neuron) for the processed fraction of a record begins. Discriminant function - the distance between the vector of the synaptic weights and discrete time series - is calculated using the DTW algorithm. Thus the neuron with the least total distance becomes the winner.

The winner neuron is located in the centre of the topological neighbourhood of co-operating neurons. Let us define lateral distance between the winner neuron (i) and and re-excited neuron (j), as $ld_{j,i}$. Topological neighbourhood $h_{j,i}$ is symmetric with regard to the point of maximum defined at $ld_{j,i} = 0$. The amplitude of the topological neighbourhood $h_{j,i}$ decreases monotonically with the increase of lateral distance $ld_{j,i}$, which is the necessary condition of neural network convergence [7]. Usually a Gaussian function if used for $h_{j,i}$ calculation (formula 5).

$$h_{j,i(d)} = \exp \left(-\frac{ld_{j,i}^2}{2 \cdot \sigma^2(n)} \right) \quad . \tag{5}$$

A decrease in the topological neighbourhood is gained at the expense of subsequent lessening the width of σ function of the topological neighbourhood $h_{j,i}$. One of possible kinds of σ value dependence on discrete time n is an exponential decline (formula 6).

$$\sigma(n) = \sigma_0 \cdot \exp \left(-\frac{n}{\tau_1} \right) \quad n = 0, 1, 2, \ldots \quad , \tag{6}$$

where σ_0 is the beginning value of σ; τ_1 - some time constant, such as the number of learning cycles.

To ensure the process of self-organisation, the synaptic weights of a neuron has to change in accordance with the input data, i.e. adapt to the input space. Let us assume that $w_j(n)$ is the vector of synaptic weights of neuron j at time moment (iteration, cycle) n. In this case, at time instant $n + 1$ the renewed vector $w_j(n+1)$ is calculated by formula (7).

$$w_j(n+1) = w_j(n) + \eta(n) \cdot h_{j,i(d)}(n) \cdot (d - w_j(n)) \ , \tag{7}$$

where η - learning rate parameter; d - discrete time series from learning dataset.

Note how the difference between discrete time series and the vector of synaptic weights is calculated in expression (7). When the load is $q = 1$, that is when each neural network is processing discrete time series with a certain fixed duration, and DTW is not used, the difference between d and $w_j(n)$ is calculated as the difference between vectors of equal length. In other cases when DTW is employed, the fact of time warping has to be taken into account. For this purpose, during the organization process the Data Mining Agent fixes in memory a warping path on whose basis the distance between the vector of synaptic weights of the winner neuron and a discrete time series was calculated. Thus the following information is becoming available: according to which value of the discrete time series the corresponding synaptic weight of the neuron has to be adjusted.

The network organization contains two main processes - initial organization followed by a convergence process. The initial organisation takes about 1000 iterations [7]. During this process each network gets an initial organization, the learning rate parameter decreases from 0.1, but remains above 0.01.

The number of iterations the convergence process takes is at least 500 times larger than the number of neurons in the network [7]. The main difference from the initial organization is that during the convergence process the topological neighbourhood of a winner neuron contains only the closest neighbours or just the winner neuron.

When the networks in the Neural Block are organized the Data Mining Agent launches the cluster formation process. At the end of this process the knowledge base will be created. During the cluster formation process each record d of the dataset that was used for training is passed to the Neural Block with imitation of On-line data flowing. Algorithms for network m_i and winner neuron n_j^* determination fully coincide with those used in network organization.

In parallel, for each neuron n_j^* these statistics are kept: records with which value of the key parameter p (transition point) have got to neuron n_j^* and how many times. Cluster c_i is a neuron n_j that at least once became the winner neuron during cluster formation. The knowledge base will contain the clusters in each neural network, as also the gathered statistics for the key parameter p in each cluster.

When the Data Management Agent changes the command to "Update", the records in the forwarded dataset will be used to update synaptic weights of neurons in the neural networks. If all record have durations that can be processed with neural networks in the Neural Block, then simply the training process with new records is launched. If Neural Block contains no neural network that can process durations that one or more new records have, then such neural network is created; and the training process with new records is launched. In both cases neural networks organization and then the knowledge base are updated.

System evaluation. To evaluate the precision of transition point forecasts made by the system, two criteria are employed: Mean Absolute Error - *MAE*, to evaluate the accuracy of the system and Logical Error to evaluate whether decisions made by the system are logically correct.

The Mean Absolute Error (*MAE*) is calculated using formula (8).

$$MAE = \frac{\sum_{i=1}^{k} |p_i - r|}{k} \quad i = [1, 2, \ldots, k] \ , \tag{8}$$

where k - the number of records used for testing; p_i - real value of the key parameter for record d_i; r - the value of the key parameter forecasted by the system.

Logical error provides information about the logical potential of the system. To calculate the logical error, it is necessary to define logically correct and logically incorrect decisions. As applied to the task of forecasting product life cycle phase transition period, logically correct and logically incorrect decisions could be described as follows:

1. Assume that discrete time series d has a duration equal to l_d, but the value of the key parameter - the period of product life cycle phase transition, is $p = p_d$, where $p_d > l_d$. This statement means that a real time of transition between the phases of the product life cycle has not come yet. Accordingly, logically correct decision is to forecast transition period r_d, where $r_d > l_d$. Logically incorrect decision in this case will be if $r_d \leq l_d$.
2. Assume that discrete time series d has a duration equal to l_d, but the value of the key parameter - the period of product life cycle phase transition, is $p = p_d$, where $p_d \leq l_d$. This statement gives evidence that real transition moment has already come. Accordingly, logically correct decision could be forecasting transition period r_d, where $r_d \leq l_d$. In its turn, logically incorrect decision will take place if $r_d > l_d$.

The statement that at $r_d = l_d$ transition has occurred can be considered correct as the availability of data about some period in record d shows that the period is logically finished and, consequently, the transition - if any was assumed in this period - is occurred.

Functional aspects of the Decision Analysis Agent. The Decision Analysis Agent can perform its actions either by receiving a request from a user, or in autonomous mode, with defined interval of time (at the end of each period) reporting the decision analysis results. Products that are analysed by Decision Analysis Agent are products that still are evolving on the market. The list of such products is monitored by the Data Management Agent.

Either in autonomous mode or by request from a user the Decision Analysis Agent starts the process, displayed in Figure 3. The depicted process includes three main steps - Determination of the Best Matching Cluster (BMC) for each of the evolving products; Formation of the List of Interest (LOI), the list of products for which it would be reasonable to reconsider the planning policy; Evaluation of a cyclic and non-cyclic planning policy for each product in the List of Interest. And finishes with the fourth step - Reporting the results of the evaluation to a user. Let us describe the processes hidden behind each of the main steps.

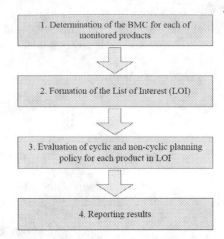

Fig. 3. Decision Analysis Agent functioning diagram

Step 1: Determination of the BMC for each of the evolving products. The Decision Analysis Agent sends a request to the Data Management Agent and receives a dataset containing evolving products. Each record is preprocessed and formatted by the Data Management Agent. As the dataset is received it being sent to the Data Mining Agent with command "Find Best Matching Cluster".

The Data Mining Agent searches the knowledge base for the Best Matching Cluster for each of the demand time series and returns a list of found clusters to the Decision Analysis Agent. The information from the BMC contains a list of possible transition points for each analysed product - for the products in the introduction phase the *M1* transition point is supplied and *M2* transition point - for products in the maturity phase. This is where the Formation of the LOI begins.

Step 2: Formation of the List of Interest. The Best Matching Cluster may contain different information for products and several cases are possible:

1. The simplest case (*C1*) when the Best Matching Cluster contains only one possible transition point. In this case the Decision Analysis Agent assumes this transition point as preferable one and follows a solution (*S1*) containing three major rules:
   (a) If $l < p$ and $p - l > \theta$ Then: Product remains monitored and is not included in the List of Interest;
   (b) If $l < p$ and $p - l \leq \theta$ Then: Product is included in the List of Interest;
   (c) If $l \geq p$ Then: Product is included in the List of Interest.
   Where l is the duration of the demand time series in periods; p - a forecasted transition point; and variable θ stores the minimal threshold of interest for either including the product in the List of Interest or not.
2. The case (*C2*) when the BMC contains more than one possible transition points for a demand time series, but one of transition points have an expressed appearance frequency f. An appearance frequency may be stated as expressed if it exceeds some threshold, like 50%. In such case the solution (*S2*) will be that the Decision

Analysis Agent accepts the transition point with an expressed f as preferable one and follows rules from the *S1* solution.

3. The third possible case (*C3*) is that the BMC contains several possible transition points, but there is no one with an expressed appearance frequency present. For this case several solutions are possible:

   (a) Solution *S3*. The Decision Analysis Agent asks the Data Mining Agent if the neural network n_i where the BMC for a current product was found, is the only one network in the Neural Block or if it is the network that processes time series with maximal possible duration. If so, then the Decision Analysis Agent follows the next two rules:

      i. If only one transition point has the highest (not expressed) f, then select it as a preferable one and follow rules from solution *S1*;

      ii. If several transition points has the highest f, then select a transition point with a minimal value as preferable one and follow rules from solution *S1*. Thus if transition points with the highest f are 4th, 5th and 8th period then the Decision Analysis Agent will choose the 4th period.

   (b) Solution *S4*. If the solution *S3* was not triggered, then the Decision Analysis Agent follows the next strategy:

      i. Decision Analysis Agent gives to the Data Mining Agent a command to find for a current product the Best Matching Clusters in all networks n_j, where $j > i$ and i is the index of the network with current BMC. That is to search for a BMC in all other networks that process time series with duration larger than the duration of a current product, excluding the current network n_i;

      ii. The Data Mining Agent searches for BMCs in the knowledge base using only first w' synaptic weights, where $w' = w_i$ and w_i is the number of synaptic weights in the current network n_i;

      iii. The list of found BMCs is returned to the Decision Analysis Agent;

      iv. Decision Analysis Agent adds the BMC from the current neural network n_i to the list, selects the most matching one cluster and checks which of three cases - *C1*, *C2* or *C3*, is triggered. If case *C1* or *C2* is triggered, then the Decision Analysis Agent just follows the rules from solutions for those cases (solutions *S1* and *S2* respectively). In case when the *C3* is triggered again, the Decision Analysis Agent follows rules from the *S3* solution.

   Example of such situation may be displayed as follows. Assume that the Neural Block contains two neural networks: first network n_1 processes time series with duration 3, 4 and 5 periods and each neuron in n_1 has 4 synaptic weights; second network n_2 is able to process time series with 6, 7 and 8 and each neuron in n_2 has 7 synaptic weights. The current product fells into the n_1 network and Decision Analysis Agent chooses to check the next networks (to use solution *S4*). In this case the BMC will be searched in the network n_2, but only first 4 of 7 synaptic weights will be used to determine the best matching cluster in the n_2.

If at the end of formation of the List of Interest the list is empty then the Decision Analysis Agent bypasses the third step and reports that products, for which it would be reasonable to reconsider the planning policy, were not found. In case when LOI contains at least one product the Decision Analysis Agent starts processes in the third step.

Step 3: Evaluation of cyclic and non-cyclic planning policy for each product in LOI. At this step the Decision Analysis Agent measures an expenses of using cyclic or non-cyclic planning policy for each product in the List of Interest. As stated in [2] the measure of Additional Cost of a Cyclic Schedule (*ACCS*) may be used for those purposes. The *ACCS* measures the gap between cyclic and non-cyclic planning policies, and is calculated by formula (9).

$$ACCS = \frac{CPPC - NCPPC}{NCPPC} , \tag{9}$$

where *CPPC* is the Cyclic Planning Policy Cost and *NCPPC* - the Non-Cyclic Planning Policy Cost.

As the third step is finished the user receives results of analysis of the products from the List of Interest.

4   Gathered Results

The fact that the data describes real life process and marks of transitions were putted by experts implies that some noisiness in data is present.

The obtained dataset contains 199 real product demand time series with minimal duration equals to 4 and maximal - to 24 periods. Each time series contains the demand during the introduction phase of a specific product plus one period of the maturity phase, and is marked with *M1* marker. To normalize the data, the Z-score with standard deviation normalization method was applied. As the true bounds of the demand data in the dataset are unknown and the difference between values of various time series is high, the chosen normalization method is one of the most suitable ones.

Figure 4 displays an example of time series data used in experiments. As can be seen, the time series differs not only in duration, but also in amplitude and its pattern.

The main target of the performed experiments was to analyse comparatively the precision of forecasting transition points with square neural network topology with 8 neighbours applied while using different network load q. The more precise the forecasted transition point will be, the more precise will be the result, returned by the Decision Analysis Agent.

The network load q was changing incrementally from one to five. To calculate the system errors - Mean Absolute Error (*MAE*) and Logical Error (*LE*), a 10-fold cross validation method was applied, totally giving 50 system runs.

Table 1 contains the size of the network, the total number of neurons, as also supplies the number of iterations for initial organization and convergence process.

The learning parameters, used for network organisation in each run, are given in Table 2. For each learning parameter the starting value and the minimal (last) value are supplied, as also the type of a function used for managing the parameter decline process.

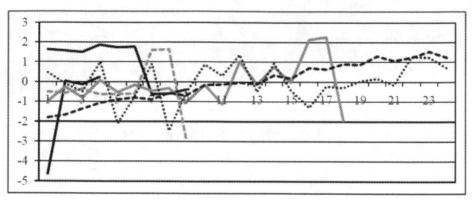

Fig. 4. Example of normalized data

Table 1. Network and learning parameters

Size	Neurons	Initial organization	Convergence process
5 x 5	25	1000 iter.	12500 iter.

Table 2. Learning coefficients

Parameter	Starts with	Ends with	Function
Learning coeff. - η	0.9	0.01	Exponential
σ for Gaussian neighbourhood	0.5	0.01	Exponential

While testing the system in On-line mode, for each of five defined values of q a Mean Absolute Error and a Logical Error were obtained. The gathered results are accumulated in Table 3, and graphically displayed in Figure 5.

The obtained results show that created agents based system with certain precision is able to predict transition points for new products, using a model, built on a historical demand data. The system was able to make a logically correct (refer to "System learning" step in subsection 3.1) decisions in at least 83.0% and at most in 87.7% of times. Thus the Mean Absolute Error lies close to 2 periods. Together with the specificity of the dataset obtained and the size of the network, it is possible to conclude that the created agents based system can be used as a data mining tool to gain an additional

Table 3. On-line Mean Absolute Error - *MAE*

Error	Topology	$q=1$	$q=2$	$q=3$	$q=4$	$q=5$
MAE	*SQR-8*	2.050	2.167	2.408	2.216	2.416
LE	*SQR-8*	13.5%	11.7%	14.5%	12.3%	17.0%

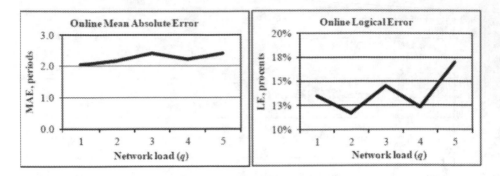

Fig. 5. Online Mean Absolute Error and Online Logical Error

knowledge for solving a planning policy management task. As well as for solving other tasks, connected with forecasting a value of a target parameter for a time dependent variable, followed by a Decision Analysis process.

5   Conclusions

For the practitioners of management of the product life cycle the knowledge, which describes in which phase the product currently is and when the transition between phases will occur, is topical. Such knowledge, in particular, helps to select between the cyclic and non-cyclic policy of planning supply chain operation.

In this paper, the task of forecasting the transition points between different phases of product life cycle is stated, and the structure of Agents Based Data Mining and Decision Support system, which helps to solve this task, is shown. The functional aspects of the Decision Analysis Agent are described. Experimentally gathered results show that the created Agents Based Data Mining driven Decision Support system has its potential and can process real demand data, create a model on the basis of historical data, forecast possible transition points and theoretically report an analysis of expenses for cyclic and non-cyclic planning policies.

For the future research it is necessary to examine the developed system on the data from different production fields, and, which is also important, to have a response from practitioners of supply chain management who will use these systems. The examination of other technologies for creating the knowledge base in the Data Mining Agent will be performed.

Another important moment is that the modest data volume that was used for practical experiments, is related to the fact, that it is necessary to have transition marks in historical data from experts and practitioners. The more products, the more complicated for human is to make all these marks - in practice the amount of marked data will always be restricted. As a result, one more possible direction of future research is treatment of forecasting the transition points in the context of a semi-supervised learning [17]. In this case, there is a small set with marked transitions and also a large dataset in which transitions are not marked. In such a situation it is necessary to create a model, which will be able to apply the knowledge, gathered on the small set of marked data, to the new (test) data.

References

1. Athanasiadis, I., Mitkas, P.: An agent-based intelligent environmental monitoring system. Management of Environmental Quality: An International Journal 15(3), 238–249 (2004)
2. Campbell, G.M., Mabert, V.A.: Cyclical schedules for capacitated lot sizing with dynamic demands. Management Science 37(4), 409–427 (1991)
3. Dunham, M.: Data Mining Introductory and Advanced Topics. Prentice Hall, Englewood Cliffs (2003)
4. Ferber, J.: Multi-Agent Systems: An Introduction to Distributed Artificial Intelligence. Pearson Education, London (1999)
5. Han, J., Kamber, M.: Data Mining: Concepts and Techniques, 2nd edn. Morgan Kaufman, San Francisco (2006)
6. Hand, D.J., Mannila, H., Smyth, P.: Principles of Data Mining. MIT Press, Cambridge (2001)
7. Haykin, S.: Neural Networks, 2nd edn. Prentice Hall, Englewood Cliffs (1999)
8. Keogh, E., Pazzani, M.: Derivative dynamic time warping. In: Proceedings of the First SIAM International Conference on Data Mining, Chicago, USA (2001)
9. Kohonen, T.: Self-Organizing Maps, 3rd edn. Springer, Heidelberg (2001)
10. Kotler, P., Armstrong, G.: Principles of Marketing, 11th edn. Prentice Hall, Englewood Cliffs (2006)
11. Liu, J.: Autonomous Agents and Multi-Agent Systems: Explorations in Learning, Self-Organization and Adaptive Computation. World Scientific, Singapore (2001)
12. Merkuryev, Y., Merkuryeva, G., Desmet, B., Jacquet-Lagreze, E.: Integrating analytical and simulation techniques in multi-echelon cyclic planning. In: Proceedings of the First Asia International Conference on Modelling and Simulation, pp. 460–464. IEEE Computer Society, Los Alamitos (2007)
13. Obermayer, K., Sejnowski, T. (eds.): Self-Organising Map Formation. MIT Press, Cambridge (2001)
14. Pyle, D.: Data Preparation for Data Mining. Morgan Kaufmann Publishers, an imprint of Elsevier, San Francisco (1999)
15. Symeonidis, A., Kehagias, D., Mitkas, P.: Intelligent policy recommendations on enterprise resource planning by the use of agent technology and data mining techniques. Expert Systems with Applications 25(4), 589–602 (2003)
16. Tan, P.-N., Steinbach, M., Kumar, V.: Introduction to Data Mining. Pearson Education, London (2006)
17. Zhu, X.: Semi-supervised learning literature survey. Technical Report 1530, Department of Computer Sciences, University of Wisconsin (2008)

Part II

Agent-Driven Data Mining

Agent-Enriched Data Mining Using an Extendable Framework

Kamal Ali Albashiri and Frans Coenen

Department of Computer Science, The University of Liverpool,
Ashton Building, Ashton Street, Liverpool L69 3BX, United Kingdom
{ali,frans}@csc.liv.ac.uk

Abstract. This paper commences with a discussion of the advantages that Multi-Agent Systems (MAS) can bring to the domain of Knowledge Discovery in Data (KDD), and presents a rational for Agent-Enriched Data Mining (AEDM). A particular challenge of any generic, general purpose, AEDM system is the extensive scope of KDD. To address this challenge the authors suggest that any truly generic AEDM must be readily extendable and propose EMADS, The Extendable Multi-Agent Data mining System. A complete overview of the architecture and agent interaction models of EMADS is presented. The system's operation is described and illustrated in terms of two KDD scenarios: meta association rule mining and classifier generation. In conclusion the authors suggest that EMADS provides a sound foundation for both KDD research and application based AEDM.

Keywords: Agent-Enriched Data Mining (AEDM), Classifier Generation, Meta Association Rule Mining (Meta ARM).

1 Motivation and Goals

Agent-Enriched Data Mining (AEDM), also known as multi-agent data mining, seeks to harness the general advantageous of MAS in the application domain of Data Mining (DM). MAS technology has much to offer DM, particularly in the context of various forms of distributed and cooperative DM. Distributed (and parallel) DM is directed at reducing the time complexity of computation associated with the increasing sophistication, size and availability of the data sets we wish to mine. Cooperative DM encompasses ensemble mechanisms and techniques such as bagging and boosting. MAS have a clear role in both these areas. MAS technology also offers some further advantageous for AEDM, namely:

- Extendibility of DM frameworks,
- Resource and experience sharing,
- Greater end-user accessibility,
- Information hiding, and
- The addressing of privacy and security issues.

The last of the above advantageous merits some further comment. By its nature DM is often applied to sensitive data. The MAS approach would allow data to be mined

L. Cao et al. (Eds.): ADMI 2009, LNCS 5680, pp. 53–68, 2009.

remotely. Similarly, with respect to DM algorithms, MAS can make use of algorithms without necessitating their transfer to users, thus contributing to the preservation of intellectual property rights. MAS make it possible for software services to be provided through the cooperative efforts of distributed collections of autonomous agents. Communication and cooperation between agents are brokered by one or more facilitators, which are responsible for matching requests, from users and agents, with descriptions of the capabilities of other agents. Thus, it is not generally required that a user or agent know the identities, locations, or number of other agents involved in satisfying a request.

The challenge of generic AEDM is the disparate nature and variety of modern DM, and the necessary communication mechanism required to cope with this disparate nature. One approach is to make use of the established Agent Communication Languages (ACLs) and mechanisms; well known examples include the Knowledge Query and Manipulation Language (KQML), the Knowledge Interchange Format (KIF), and the Foundation for Intelligent Physical Agents (FIPA) ACL [14]. All these ACLs have their advantageous and disadvantageous and tend to address particular forms of intra-agent communication; for example FIPA ACL is directed at agent negotiation. Each can be employed in the context of AEDM communication but on its own will not facilitate the shared agent understanding required to achieve generic AEDM. This would require recourse to the use of ontologies and/or some agreed meta-language. It is suggested in this work that a method of addressing the communication requirements of AEDM is to use a system of mediators and wrappers coupled with an ACL such as FIPA ACL, and that this can more readily address the issues concerned with the variety and range of contexts to which AEDM can be applicable.

To investigate and evaluate the expected advantageous of wrappers and mediators, in the context of generic AEDM, the authors have developed and implemented (in JADE) a multi-agent platform, EMADS (the Extendable Multi-Agent Data mining System). Extendibility is seen as an essential feature of the framework primarily because it allows its functionality to grow in an incremental manner. The vision is of an "anarchic" collection of agents, contributed to by a community of EMADS users, that exist across an "internet space"; that can negotiate with each other to attempt to perform a variety of DM tasks (or not if no suitable collection of agents come together) as proposed by other (or the same) EMADS users. An EMADS demonstrator is currently in operation.

The primary goal of the EMADS framework is to provide a means for integrating new DM algorithms and data sources in a distributed infrastructure and collaborative environment. However, EMADS also seeks to address some of the issues of DM that would benefit from the rich and complex interactions of communicating agents. The broad advantages offered by the framework are:

– Flexibility in assembling communities of autonomous service providers, including the incorporation of existing applications.
– Minimization of the effort required to create new agents, and to wrap existing applications.
– Support for end users to express DM requests without having detailed knowledge of the individual agents.

The rest of this paper is organised as follows. A brief review of some related work on Agent-enriched Data Mining (AEDM) is presented in Section 2. The conceptual framework together with an overview of the wrapper principle is presented in Section 3. The framework operation is illustrated in Section 4 using two DM scenarios: Meta Association Rule Mining (MARM) and single label classification. Finally some conclusions are presented in Section 5.

2 Related Work

There are a number of reports in the literature of the application of Agent techniques to DM. The contribution of this section is a broad review of prominent AEDM approaches in the literature and discussion of the benefits that agent-driven DM architectures provide in coping with such problems. This section is not concerned with particular DM techniques; it is however concerned with work on the design of distributed and multi-agent system directed at DM.

The most fundamental approach to distributed DM is to move all of the data to a central data warehouse and then to analyze this with a single DM system, even though this approach intuitively guarantees accurate DM results, it might be infeasible in many cases.

An alternative approach is high level learning with meta-learning strategies in which all the data can be locally analyzed (local data model), and the local results at their local sites combined at the central site to obtain the final result (global data model). Meta-learning methods have been widely used within DM [30,10], particularly in the area of classification and regression. These approaches are less expensive but may produce ambiguous and incorrect global results. In addition, Distributed DM approaches require centralised control that causes a communication bottleneck that sometimes leads, in turn, to inefficient performance and system failure.

To make up for such a weakness, many researchers have investigated more advanced approaches of combining local models built at different sites. Most of these approaches are agent-based high level learning strategies.

One of the earliest references to AEDM can be found in Kargupta et al. [19] who describe a parallel DM system (PADMA) that uses software agents for local data accessing and analysis, and a web based interface for interactive data visualization. PADMA has been used in medical applications. They describe a distributed DM architecture and a set of protocols for a multi-agent software tool. Peng et al. [25] presented an interesting comparison between single-agent and multi-agent text classification in terms of a number of criteria including response time, quality of classification, and economic/privacy considerations. Their results indicate, not unexpectedly, in favour of a multi-agent approach.

A popular AEDM approach is described in the METAL project [24] whose emphasis is on helping the user to obtain a ranking of suitable DM algorithms through an online advisory system. Gorodetsky et al. [16] correctly consider that the core problem in AEDM is not the DM algorithms themselves (in many case these are well understood), but the most appropriate mechanisms to allow agents to collaborate. Gorodetsky et al. present an AEDM system to achieve distributed DM and, specifically, classification. A more recent system, proposed in [23], uses the MAGE middleware [28] to build an execution engine that uses a directed acyclic graph to formalize the representation of

KDD process. In [11] a multi-agent system F-Trade has been proposed. It is a web-based DM infrastructure for trading and surveillance support in capital markets.

The meta-learning strategy offers a way to mine classifiers from homogeneously distributed data. It follows three main steps. The first is to generate base classifiers at each site using a classifier learning algorithms. The second step is to collect the base classifiers at a central site, and produce meta-level data from a separate validation set and predictions generated by the base classifier on it. The third step is to generate the final classifier (meta-classifier) from meta-level data via a combiner or an arbiter. Copies of the classifier agent will exist, or be deployed, on nodes in the network being used (see for example [26]). Perhaps the most mature agent-based meta-learning systems are: JAM [29], BODHI [18], and Papyrus [6]. Papyrus is designed to support both learning strategies; meat-learning and central learning. A hybrid learning strategy is a technique that combines local and remote learning for model building [17]. In contrast to JAM and BODHI, Papyrus can not only move models from site to site, but can also move data when such a strategy is desirable. Papyrus is a specialized system which is designed for clusters while JAM and BODHI are designed for data classification. These are reviewed in details in [20].

Most of the previously proposed AEDM systems are used to improve the performance of one specific DM task. To the best knowledge of the authors, there have been only few AEDM systems that define a generic framework for the AEDM approach. An early attempt was IDM [9], a multiple agent architecture that attempts to do direct DM that helps businesses gather intelligence about their internal commerce agent heuristics and architectures for KDD. In [5] a generic task framework was introduced but was designed to work only with spatial data. The most recent system is introduced in [15] where the authors proposed a multi-agent system to provide a general framework for distributed DM applications. The effort to embed the logic of a specific domain has been minimized and is limited to the customization of the user. However, although its customizable feature is of a considerable benefit, it still requires users to have very good DM knowledge.

3   EMADS Overview

A high level view of the framework conceptualization showing the various categories of agents and their contributors is given in Figure 1. The housekeeping (DF and AMS) agents are specialized server agents that are responsible for helping agents to locate one another. They do not participate in problem-solving; they only play a role of a facilitator in the system. Note that any system configuration is not limited to single MAS. Larger systems can be assembled from multiple MASs, each having the sort of structure shown in Figure 2.

3.1   System Structure

The EMADS framework has several different modes of operation according to the nature of the participant. Each mode of operation has a corresponding category of *User Agent*. Broadly, the supported categories are:

- Developers: Developers are participants, who have full access and may contribute DM algorithms in the form of *Data Mining Agents* (DM Agents).
- Data Miners: These are participants, with restricted access to the system, who may pose DM requests through User Agents and *Task Agents* (see below for further details).
- Data Contributors: These are participants, again with restricted access, who are prepared to make data available, by launching *Data Agents*, to be used by DM agents.

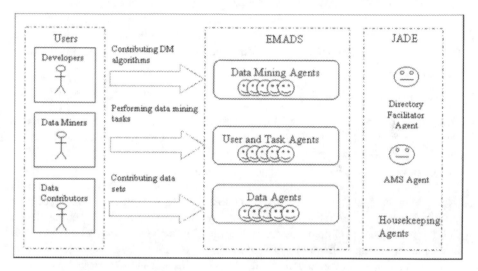

Fig. 1. High level view of EMADS conceptual framework

The various categories of agents are illustrated in Figure 1: DM agents, Task Agents, User Agents and Data Agents. DM agents are usually specialist agents that posses an algorithm for a particular DM task or sub-task. DM agents may be based on legacy applications, in which case the agent may be little more than a wrapper that calls a pre-existing API (see subsection 3.3 for further detail).

Note that, before interaction with EMADS can commence, appropriate software needs to be downloaded and launched by the participant. Note also that any individual participant may be as a user, contributor and developer at the same time.

Conceptually the nature of the requests that may be posted by users is extensive. In the current demonstration implementation, the following types of generic request are supported:

- Find the "best" classifier (to be used by the requester at some later date in off line mode) for a data set provided by the user.
- Find the "best" classifier for the indicated data set (i.e. provided by some other participant).
- Find a set of Association Rules (ARs) contained within the data set(s) provided by the user.
- Find a set of Association Rules (ARs) contained within the indicated type of data set(s) (i.e. provided by other participants).

In the above a "best" classifier is defined as a classifier that will produce the highest accuracy on a given test set (identified by the mining agent) according to the detail of the request. To obtain the "best" classifier EMADS will attempt to access and communicate with as many classifier generator DM agents as possible and select the best result. The classification style of user request will be discussed further in subsection 4.2 to illustrate the operation of EMADS in more detail.

The Association Rule Mining (ARM) style of request is discussed further in subsection 4.1. The scenario investigated here is one where an agent framework is used to implement a form of Meta-ARM where the results of the parallel application of ARM to a collection of data sets, with not necessarily the same schema but conforming to a global schema, are combined. Some further details of this process can be found in Albashiri et al. [3,4].

3.2 Agent Interactions

Conceptually the EMADS system is a hybrid peer to peer agent based system comprising a collection of collaborating agents that exist in a set of containers. Agents may be created and contributed to the system by any user/contributor. One of these containers, the main container, holds a number of housekeeping agents that provide various facilities to maintain the operation of the framework. In particular the main container holds an Agent Management System (AMS) agent and a Directory Facilitator (DF) agent. The terminology used is taken from the JADE (Java Agent Development) [7] framework in which the framework is implemented. Briefly the AMS agent is used to control the life cycles of other agents in the platform, and the DF agent provides an agent lookup service. Both the main container and the remaining containers can hold various DM agents. Note that the EMADS main container is located on the EMADS host organisation site, while the other containers may be held at any other sites worldwide.

Figure 2 presents the EMADS architecture as implemented in JADE. It shows a sample collection of several application agents and housekeeping agents, organized as a community of peers by a common relationship to each other.

With reference to Figure 2, a user agent runs on the user's local host and is responsible for: (i) accepting user input (request), (ii) launching the appropriate Task Agent to processes the user request, and (iii) displaying the results of the (distributed) computation. The user expresses a task to be executed using standard interface dialogue mechanisms by clicking on active areas in the interface and, in some cases, by entering thresholds values; note that the user does not need to specify which agent or agents should be employed to perform the desired task. For instance, if the question "What is the best classifier for my data?" is posed in the user interface, this request will trigger a Task Agent. The Task Agent requests the facilitator to match the action part of the request to capabilities published by other agents. The request is then routed by the Task Agent to the appropriate agents to execute the request, this will typically involve communication among various relevant agents within the system. On completion the results are sent back to the user agent for display.

The key elements of the operation of EMADS that should be noted are:

1. The mechanism whereby a collection of agents can be harnessed to identify a "best solution".

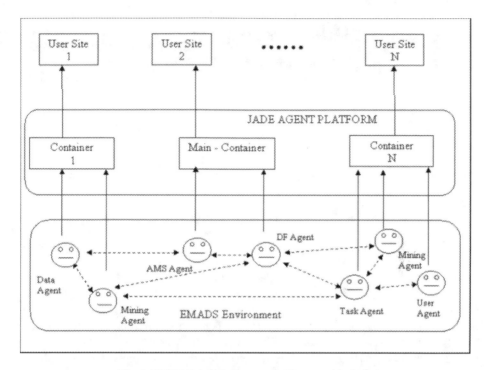

Fig. 2. EMADS Architecture as Implemented in Jade

2. The process whereby new agents connect to the facilitator and registering their capability specifications.
3. That the interpretation and execution of a task is a distributed process, with no one agent defining the set of possible inputs to the system.
4. That a single request can produce cooperation and flexible communication among many agents spread across multiple machines.

Agent Cooperation. Cooperation among the various EMADS agents is achieved via messages expressed in FIPA ACL and is normally structured around a three-stage process:

1. **Service Registration** where providers (agents who wish to provide *services*) register their capability specifications with a facilitator.
2. **Request Posting** where User Agents (*requesters* of services) construct requests and relay them to a Task Agent,
3. **Processing** where the Task Agent coordinates the efforts of the appropriate service providers (Data Agents and DM Agents) to satisfy the request.

Note that Stage 1 (service registration) is not necessarily immediately followed by stage 2 and 3, it is possible that a provider services may never be used. Note also that the facilitator (the DF and AMS agents) maintains a knowledge base that records the capabilities of the various EMADS agents, and uses this knowledge to assist requesters and providers of services in making contact. When a service provider (i.e. Data Agent or DM Agent) is created, it makes a connection to a facilitator, which is known as its *parent facilitator*. Upon connection, the new agent informs its parent facilitator of the services it can provide. When the agent is needed, the facilitator sends its address to the requester agent. An important element of the desired EMADS agent cooperation model is the function of the Task Agent; this is therefore described in more detail in the following subsection.

The Task Agent. A Task Agent is designed to handle a user request. This involves a three step process:

1. Determination of whom (which specific agents) will execute a request;
2. Optimization of the complete task, including parallelization where appropriate; and
3. Interpretation of the optimized task.

Thus determination (step 1) involves the selection of one or more agents to handle each sub-task given a particular request. In doing this, the Task agent uses the facilitator's knowledge of the capabilities of the available EMADS agents (and possibly of other facilitators, in a multi-facilitator system). The facilitator may also use information specified by the user (such as threshold values). In processing a request, an agent can also make use of a variety of capabilities provided by other agents. For example, an agent can request data from Data Agents that maintain shared data. The optimization step results in a request whose interpretation will require as few communication exchanges as possible, between the Task Agent and the satisfying agents (typically DM Agents and Data Agents), and can exploit the parallel processing capabilities of the satisfying agents. Thus, in summary, the interpretation of a task by a Task Agent involves: (i) the coordination of requests directed at the satisfying agents, and (ii) assembling the responses into a coherent whole, for return to the user agent.

3.3 System Extendibility

One of the principal objectives of the EMADS framework is to provide an easily extendable framework that can accept new data sources and new DM techniques. In general, extendibility can be defined as the ease with which software can be modified to adapt to new requirements or changes in existing requirements. Adding a new data source or DM techniques should be as easy as adding new agents to the system. The desired extendibility is achieved by a system of wrappers. EMADS wrappers are used to "wrap" up DM artifacts so that they become EMADS agents and can communicate with other EMADS agents. Such EMADS wrappers can be viewed as agents in their own right that are subsumed once they have been integrated with data or tools to become EMADS agents. The wrappers essentially provide an application interface to EMADS that has to be implemented by the end user, although this has been designed to be a fairly trivial operation.

In the current demonstration EMADS system two broad categories of wrapper have been defined: (i) data wrappers and (ii) tool wrappers; the first is used to create data agents and the second to create DM agents. Each is described in further detail in the following two subsections.

Data Wrappers. Data wrappers are used to "wrap" a data source and consequently create a data agent. Broadly a data wrapper holds the location (file path) of a data source, so that it can be accessed by other agents; and meta information about the data. To assist end users in the application of data wrappers a data wrapper GUI is available. Once created, the data agent announces itself to the DF agent as a consequence of which it becomes available to all EMADS users.

Tool Wrappers. Tool wrappers are used to "wrap" up DM software systems and thus create a mining agent. Generally the software systems will be DM tools of various kinds (classifiers, clusters, association rule miners, etc.) although they could also be (say) data normalization/discretization or visualization tools. It is intended that the framework will incorporate a substantial number of different tool wrappers each defined by the nature of the desired I/O which in turn will be informed by the nature of the generic DM tasks that it is desirable for EMADS to be able to perform. Currently the research team has implemented two tool wrappers:

1. The binary valued data, single label, classifier generator.
2. The data normalization/discretization wrapper.

Many more categories of tool wrapper can be envisaged. Mining tool wrappers are more complex than data wrappers because of the different kinds of information that needs to be exchanged.

In the case of a "binary valued, single label, classifier generator" wrapper the input is a binary valued data set together with meta information about the number of classes and a number slots to allow for the (optional) inclusion of threshold values. The output is then a classifier expressed as a set of Classification Rules (CRs). As with data agents, once created, the DM agent announce themselves to the DF agent after which they will becomes available for use to EMADS users.

For example, in the case of the data normalization/discretization wrapper, the LUCS-KDD (Liverpool University Computer Science - Knowledge Discovery in Data) ARM DN (Discretization/ Normalization) software[1] is used to convert data files, such as those available in the UCI data repository [8], into a binary format suitable for use with Association Rule Mining (ARM) applications. This tool has been "wrapped" using the data normalization/discretization wrapper.

4   System Demonstration

Perhaps the best way to obtain an intuitive sense of how the framework typically functions is to briefly look at an example of how it has been applied to real world scenarios.

---

[1] $http://www.csc.liv.ac.uk/\tilde{f}rans/KDD/Software/$

The following subsections describe two demonstration applications (Scenarios) implemented within the EMADS framework.

The first (discussed further in subsection 4.1) is a distributed meta mining scenario where EMADS agents are used to merge the results of a number of ARM operations, a process referred to as meta-ARM, to produce a global set of Association Rules (ARs). The challenge here is to minimise the communication overhead, a significant issue in distributed and parallel DM (regardless of whether it is implemented in an agent framework or not).

The second scenario (subsection 4.2) is a classification scenario where the objective is to generate a classifier (predictor) fitted to EMADS user's specified data set. It has been well established within the DM research community, for reasons that remain unclear but are concerned with the nature of the input data, that there is no single "best" classification algorithm. The aim of this second scenario is therefore to identify a "best" classifier given a particular data set. Best in this context is measured in terms of classification accuracy. This experiment not only serves to illustrate the advantageous of EMADS but also provides an interesting comparison of a variety of classification techniques and algorithms.

4.1 Meta ARM (Association Rule Mining) Scenario

The term *meta mining* is defined, in the context of EMADS, as the process of combining the individually obtained results of N applications of a DM activity. The motivation behind the scenario is that data relevant to a particular DM application may be owned and maintained by different, geographically dispersed, organizations. There is therefore a "privacy and security" issue, privacy preserving issues [1] are of major concerns in inter enterprise DM when dealing with private databases located at different sites. One approach to addressing the meta mining problem is to adopt a distributed approach. The meta mining scenario considered here is a meta Association Rule Mining (meta ARM) scenario where the results of N ARM operations, by N agents, are brought together.

Dynamic Behaviour of System for Meta ARM Operations. The meta ARM EMADS illustration described here was used to identify the most efficient Meta ARM agent process given a number of alternatives. The first algorithm was a bench mark algorithm, against which other Meta ARM algorithms were compared. Four comparison meta ARM algorithms were constructed (Apriori, Brute Force, Hybrid 1 and Hybrid 2). Full details of the algorithms can be found in [3]. In each case it was assumed that each data source would produce a set of frequent sets, using some ARM algorithm, with the results stored in a common data structure. These data structures would then be merged in some manner through a process of agent collaboration. Each of the Meta ARM algorithms made use of a Return To Data (RTD) lists, one per data set, to contain lists of itemsets whose support was not included in the original ARM operation and for which the count was to be obtained by a return to the raw data held at a data agent. The RTD lists comprised zero, one or more tuples of the form $< I, sup >$, where I is an item set for which a count is required and sup is the desired count. RTD lists are constructed as a meta ARM algorithm progresses. During RTD list construction the sup value will be 0, it is not until the RTD list is processed that actual values are assigned to sup. The processing of RTD lists may occur during, and/or at the end of, the meta ARM process depending on the nature of the meta ARM algorithm used.

The meta ARM scenario comprises a set of N data agents and $N + 1$ DM agents: N ARM agents and one meta ARM agent. Note that each ARM agent could have a different ARM algorithm associated with it, however a common data structure was assumed to facilitate data interchange. The common data structure used was a T-tree [12], a set enumeration tree structure for storing item sets. Once generated the N local T-trees were passed to the Meta ARM agent which created a global T-tree. During the global T-tree generation process the Meta ARM agent interacted with the various ARM agents in the form of the exchange of RTD lists.

Experimentation and Analysis. To evaluate the five Meta ARM algorithms (including the bench mark algorithm), in the context of the EMADS vision, a number of experiments were conducted. These are described and analyzed in this subsection. The experiments were designed to analyze the effect of the following:

1. The number of data sources (data agents).
2. The size of the data sets (held at data agents) in terms of number of records.
3. The size of the data sets (held at data agents) in terms of number of attributes.

Experiments were run using two Intel Core 2 Duo E6400 CPU (2.13GHz) computers with 3GB of main memory (DDR2 800MHz), Fedora Core 6, Kernel version 2.6.18 running under Linux except for the first experiment where two further computers running under Windows XP were added. For each of the experiments we measured:

- Processing time (seconds/mseconds),
- The size of the RTD lists (Kbytes), and
- The number of RTD lists generated.

Note that the authors did not use the well known IBM QUEST generator [2] because many different data sets (with the same input parameters) were required and it was found that the QUEST generator always generated the same data given the same input parameters. Instead the authors used the LUCS KDD data generator[2]. Figure 3 shows the effect of adding additional data sources using the four Meta ARM algorithms and the bench mark algorithm. For this experiment thirteen different artificial data sets were generated and distributed among four machines using $T = 4$ (average number of items per transactions), $N = 20$ (Number of attributes), $D = 100k$ (Number of transactions. Note that the slight oscillations in the graphs result simply from a vagary of the random nature of the test data generation.

Figure 3 also indicate, at least with respect to meta ARM, that EMADS offers positive advantages in that all the Meta ARM algorithms were more computationally efficient than the bench mark algorithm. The results of the analysis also indicated that the Apriori Meta ARM approach coped best with a large number of data sources, while the Brute Force and Hybrid 1 approaches coped best with increased data sizes (in terms of column/rows).

---

[2] $http : //www.csc.liv.ac.uk/\tilde{f}rans/KDD/Software//LUCS - KDD - DataGen/$

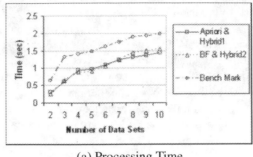

(a) Processing Time

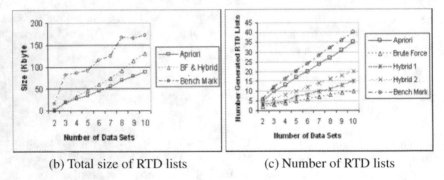

(b) Total size of RTD lists   (c) Number of RTD lists

Fig. 3. Effect of number of data sources

4.2   Classifier Generation Scenario

In this subsection the operation of EMADS is illustrated in the context of a classifier generation task, however much of the discussion is equally applicable to other generic DM tasks. The scenario is that of an end user who wishes to obtain a "best" classifier founded on a given, pre-labelled, data set; which can then be applied to further un-labelled data. The assumption is that the given data set is binary valued and that the user requires a single-label, as opposed to a multi-labelled, classifier. The request is made using the individual's user agent which in turn will spawn an appropriate task agent.

For this scenario the task agent interacts with mining agents that hold single labelled classifier generators that take binary valued data as input. Each of these mining agents is then accessed and a classifier, together with an accuracy estimate, requested. Once received the task agent selects the classifier with the best accuracy and returns this to the user agent.

The DM agent wrapper in this case provides the interface that allows input of: (i) the identifier for the data set to be classified, and (ii) the number of class attributes (a value that the mining agent cannot currently deduce for itself); while the user agent interface allows input for threshold values (such as support and confidence values).

The output is a classifier together with an accuracy measure. To obtain the accuracy measures the classifier generators (DM agents) build their individual classifier using the first half of the input data as the "training" set and the second half of the data as the "test" set. An alternative approach might have been to use Ten Cross Validation (TCV) to identify the best accuracy. It should be noted that the objective here is to return a classifier, using TCV ten classifiers will be built and thus one of them would have to be selected.

From the literature there are many reported techniques available for generating classifiers. For the scenario reported here the authors' used implementations of eight different algorithms[3]:

1. FOIL (First Order Inductive Learner) [27]: The well established inductive learning algorithm for the generation of Classification Association Rules (CARs).
2. TFPC (Total From Partial Classification): A CAR generator [13] founded on the P and T-tree set enumeration tree data structures.
3. PRM (Predictive Rule Mining) [31]: An extension of FOIL.
4. CPAR (Classification based on Predictive Association Rules) [31]: A further development from FOIL and PRM.
5. IGDT (Information Gain Decision Tree) classifier: An implementation of the well established C4.5 style of decision tree based classifier using information gain as the "splitting criteria".
6. RDT (Random Decision Tree) classifier: A decision tree based classifier that uses most frequent current attribute as the "splitting criteria".
7. CMAR (Classification based on Multiple Association Rules): A well established Classification Association Rule Mining (CARM) algorithm [22].
8. CBA (Classification Based on Associations): Another well established CARM algorithm [21].

These were placed within an appropriately defined tool wrapper to produce eight (single label binary data classifier generator) DM agents. This was found to be a trivial operation indicating the versatility of the wrapper concept.

Thus each mining agent's basic function was to generate a classification model using its own classifier and provide this to the task agent. The task agent then evaluates all the classifier models and chooses the most accurate model to be returned to the user agent.

Experimentation and Analysis. To evaluate the EMADS classification scenario, as described above, a sub-set of the data sets available at the UCI machine learning data repository [8] were used (preprocessed by data agents so that they were discretized /normalized into a binary form). The results are presented in Table 1. Each row in the table represents a particular request and gives the name of the data set, the selected best algorithm as identified from the interaction between the EMADS agents, the resulting best accuracy and the total EMADS execution time from creation of the initial Task Agent to the final "best" classifier being returned to the user agent. The naming convention used in the Table is that: D equals the number of attributes (after discretization/normalization), N the number of records and C the number of classes (although EMADS has no requirement for the adoption of this convention).

---

[3] Taken from the LUCS-KDD repository at $http://www.csc.liv.ac.uk/\tilde{f}rans/KDD/Software/$

Table 1. Classification Results

Data Set	Classifier	Accuracy	Generation Time (sec)
connect4.D129.N67557.C3	RDT	79.76	502.65
adult.D97.N48842.C2	IGDT	86.05	86.17
letRecog.D106.N20000.C26	RDT	91.79	31.52
anneal.D73.N898.C6	FOIL	98.44	5.82
breast.D20.N699.C2	IGDT	93.98	1.28
congres.D34.N435.C2	RDT	100	3.69
cylBands.D124.N540.C2	RDT	97.78	41.9
dematology.D49.N366.C6	RDT	96.17	11.28
heart.D52.N303.C5	RDT	96.02	3.04
auto.D137.N205.C7	IGDT	76.47	12.17
penDigits.D89.N10992.C10	RDT	99.18	13.77
soybean.D118.N683.C19	RDT	98.83	13.22
waveform.D101.N5000.C3	RDT	96.81	11.97

The results demonstrate firstly, that EMADS can usefully be adopted to produce a best classifier from a selection of classifiers. Secondly, that the operation of EMADS is not significantly hindered by agent communication overheads, although this has some effect. Generation time, in most cases does not seem to be an issue, so further classifier generator mining agents could easily be added. The results also reinforce the often observed phenomena that there is no single best classifier generator suited to all kinds of data.

5   Conclusions

This paper described EMADS, a multi-agent framework for DM. The architecture provides a framework for the construction and operation of distributed software agents. The principal advantages offered by the system are that of experience and resource sharing, flexibility and extendibility, and (to an extent) protection of privacy and intellectual property rights. The use of a single facilitator offers both advantages and weaknesses with respect to scalability and fault tolerance. On the plus side, the grouping of a facilitator with a collection of agents provides a faster look-up service. However, even though the intention is that the facilitator assists agents in finding one another and then to "step aside" while other agents communicate over a direct, dedicated channel so as to prevent a communication bottleneck; there is still the potential for a facilitator to become a critical point of system failure. Further work in this area is therefore required, one solution is to use more than one facilitator deployed on multiple machines for a better fault-tolerant platform.

The framework's operation is illustrated using both meta ARM and classification scenarios. Extendibility is presented by showing how wrappers are used to incorporate existing software into EMADS. Experience to date indicates that, given an appropriate wrapper, existing DM software can very easily be packaged to become a DM agent.

Flexibility is illustrated using a classification scenario. Information hiding is illustrated in that users need have no knowledge of how any particular piece of DM software works or the location of the data to be used.

A good foundation has been established for both DM research and genuine application based DM. It is acknowledged that the current functionality of the framework is limited to classification and meta ARM. The research team is at present working towards increasing the diversity of mining tasks that can be addressed. There are many directions in which the work can (and is being) taken forward. One interesting direction is to build on the wealth of distributed DM research that is currently available and progress this in an MAS context. The research team is also enhancing the system's robustness so as to make it publicly available. It is hoped that once the system is live other interested DM practitioners will be prepared to contribute algorithms and data.

References

1. Aggarwal, C., Yu, P.: A Condensation Approach to Privacy Preserving DataMining. In: Bertino, E., Christodoulakis, S., Plexousakis, D., Christophides, V., Koubarakis, M., Böhm, K., Ferrari, E. (eds.) EDBT 2004. LNCS, vol. 2992, pp. 183–199. Springer, Heidelberg (2004)

2. Agrawal, R., Mehta, M., Shafer, J., Srikant, R., Arning, A., Bollinger, T.: The Quest Data Mining System. In: Proceedings 2nd Int. Conf. Knowledge Discovery and Data Mining, KDD (1996)

3. Albashiri, K., Coenen, F., Sanderson, R., Leng, P.: Frequent Set Meta Mining: Towards Multi-Agent Data Mining. In: Bramer, M., Coenen, F.P., Petridis, M. (eds.) Research and Development in Intelligent Systems XXIII, pp. 139–151. Springer, London (2007)

4. Albashiri, K., Coenen, F., Leng, P.: Agent Based Frequent Set Meta Mining: Introducing EMADS. In: Artificial Intelligence in Theory and Practice II, Proceedings IFIP, pp. 23–32. Springer, Heidelberg (2007)

5. Baazaoui, H., Faiz, S., Ben Hamed, R., Ben Ghezala, H.: A Framework for data mining based multi-agent: an application to spatial data. In: 3rd World Enformatika Conference, WEC 2005, Avril, Istanbul (2005)

6. Bailey, S., Grossman, R., Sivakumar, H., Turinsky, A.: Papyrus: a system for data mining over local and wide area clusters and super-clusters. In: Proceedings Conference on Supercomputing, p. 63. ACM Press, New York (1999)

7. Bellifemine, F., Poggi, A., Rimassi, G.: JADE: A FIPA-Compliant agent framework. In: Proceedings Practical Applications of Intelligent Agents and Multi-Agents, pp. 97–108 (1999), http://sharon.cselt.it/projects/jade

8. Blake, C., Merz, C.: UCI Repository of machine learning databases. University of California, Department of Information and Computer Science, Irvine, CA (1998), http://www.ics.uci.edu/mlearn/MLRepository.html

9. Bose, R., Sugumaran, V.: IDM: An Intelligent Software Agent Based DataMining Environment. In: Proceedings of the IEEE International Conference on Systems, Man, and Cybernetics, pp. 2888–2893. IEEE Press, San Diego (1998)

10. Bota, J., Gmez-Skarmeta, A., Valds, M., Padilla, A.: Metala: A meta-learning architecture. Fuzzy Days, 688–698 (2001)

11. Cao, L., Zhang, C.: F-Trade: Agent-mining symbiont for financial services. In: AAMAS, pp. 1363–1364 (2007)

12. Coenen, F., Leng, P., Goulbourne, G.: Tree Structures for Mining Association Rules. Journal of Data Mining and KDD 8, 25–51 (2004)

13. Coenen, F., Leng, P., Zhang, L.: Threshold Tuning for Improved Classification Association Rule Mining. In: Ho, T.-B., Cheung, D., Liu, H. (eds.) PAKDD 2005. LNCS, vol. 3518, pp. 216–225. Springer, Heidelberg (2005)
14. Foundation for Intelligent Physical Agents, FIPA 2002 Specification. Geneva, Switzerland (2002), http://www.fipa.org/specifications/index.html
15. Giuseppe, D., Giancarlo, F.: A customizable multi-agent system for distributed data mining. In: Proc. of the 2007 ACM symp. on applied computing, pp. 42–47 (2007)
16. Gorodetsky, V., Karsaeyv, O., Samoilov, V.: Multi-agent technology for distributed data mining and classification. In: Proceedings Int. Conf. on Intelligent Agent Technology (IAT 2003), IEEE/WIC, pp. 438–441 (2003)
17. Grossman, R., Turinsky, A.: A framework for finding distributed data mining strategies that are intermediate between centralized strategies and in-place strategies. In: KDD Workshop on Distributed Data Mining (2000)
18. Kargupta, H., Byung-Hoon, et al.: Collective Data Mining: A New Perspective Toward Distributed Data Mining. In: Advances in Distributed and Parallel Knowledge Discovery. MIT/AAAI Press (1999)
19. Kargupta, H., Hamzaoglu, I., Stafford, B.: Scalable, Distributed Data Mining Using an Agent Based Architecture. In: Proceedings of Knowledge Discovery and Data Mining, pp. 211–214. AAAI Press, Menlo Park (1997)
20. Klusch, M., Lodi, G.: Agent-based Distributed Data Mining: The KDEC Scheme. In: Klusch, M., Bergamaschi, S., Edwards, P., Petta, P. (eds.) Intelligent Information Agents. LNCS, vol. 2586, pp. 104–122. Springer, Heidelberg (2003)
21. Liu, B., Hsu, W., Ma, Y.: Integrating Classification and Assocoiation Rule Mining. In: Proceedings KDD 1998, New York, August 27-31, pp. 80–86. AAAI, Menlo Park (1998)
22. Li, W., Han, J., Pei, J.: CMAR: Accurate and Efficient Classification Based on Multiple Class-Association Rules. In: Proceedings ICDM, pp. 369–376 (2001)
23. Luo, P., Huang, R., He, Q., Lin, F., Shi, Z.: Execution engine of meta-learning system for kdd in multi-agent environment. Technical report, Institute of Computing Technology. Chinese Academy of Sciences (2005)
24. METAL Project. Esprit Project METAL (2002), http://www.metal-kdd.org
25. Peng, S., Mukhopadhyay, S., Raje, R., Palakal, M., Mostafa, J.: A Comparison Between Single-agent and Multi-agent Classification of Documents. In: Proceedings 15th Intern. PD Processing Symposium, pp. 935–944 (2001)
26. Prodromides, A., Chan, P., Stolfo, S.: Meta-Learning in Distributed Data Mining Systems: Issues and Approaches, pp. 81–114. AAAI Press/The MIT Press (2000)
27. Quinlan, J.R., Cameron-Jones, R.M.: FOIL: A Midterm Report. In: Brazdil, P.B. (ed.) ECML 1993. LNCS, vol. 667, pp. 3–20. Springer, Heidelberg (1993)
28. Shi, Z., Zhang, H., Cheng, Y., Jiang, Y., Sheng, Q., Zhao, Z.: Mage: An agent-oriented programming environment. In: Proceedings of the IEEE International Conference on Cognitive Informatics, pp. 250–257 (2004)
29. Stolfo, S., Prodromidis, A.L., Tselepis, S., Lee, W.: JAM: Java Agents for Meta-Learning over Distributed Databases. In: Proceedings of the International Conference on Knowledge Discovery and Data Mining, pp. 74–81 (1997)
30. Vilalta, R., Christophe, G., Giraud-Carrier, B.P., Soares, C.: Using Meta-Learning to Support Data Mining. IJCSA 1(1), 31–45 (2004)
31. Yin, X., Han, J.: CPAR: Classification based on Predictive Association Rules. In: Proc. SIAM Int. Conf. on Data Mining (SDM 2003), SF, CA, pp. 331–335 (2003)

Auto-Clustering Using Particle Swarm Optimization and Bacterial Foraging

Jakob R. Olesen, Jorge Cordero H., and Yifeng Zeng

Department of Computer Science, Aalborg University, Selma Lagerlfs Vej 300 Aalborg 9220, Denmark
jpcordero@exatec.itesm.mx, yfzeng@cs.aau.dk, jro@cs.aau.dk
http://www.cs.aau.dk/~yfzeng/

Abstract. This paper presents a hybrid approach for clustering based on particle swarm optimization (PSO) and bacteria foraging algorithms (BFA). The new method *AutoCPB* (Auto-Clustering based on particle bacterial foraging) makes use of autonomous agents whose primary objective is to cluster chunks of data by using simplistic collaboration. Inspired by the advances in clustering using particle swarm optimization, we suggest further improvements. Moreover, we gathered standard benchmark datasets and compared our new approach against the standard *K-means* algorithm, obtaining promising results. Our hybrid mechanism outperforms earlier PSO-based approaches by using simplistic communication between agents.

1 Introduction

How can we define a group of highly correlated genes while examining gene expression data? How can we find those relevant features in a dataset having numerous variables? How could we define a set of classes over a collection of observations? Basically, all these questions have been the fundamental motivation for one of the most popular branches of data mining: data clustering.

Clustering is one of the most important unsupervised learning problems that computer scientists, statisticians and mathematicians have tried to develop and improve for years. For more than two decades, clustering has taken special interest between the scientific community and it is probably in the jargon of other professionals that have no direct relationship with machine learning or even computer science. The reason for the later relies beneath its clear definition and interpretation.

The aim of data clustering is to recognize patterns in data and form groups (clusters C) of interdependent objects. According to [1], this task can be classified in four major paradigms:

- Model based methods.
- Hierarchical methods.
- Partitioning methods.
- Density estimation methods.

Besides of the previous classification, clustering approaches can also be independently classified in other terms. They can be either exhaustive or not exhaustive methods (exhaustive clustering associates every object to a given cluster, whereas in the second case

L. Cao et al. (Eds.): ADMI 2009, LNCS 5680, pp. 69–83, 2009.

some variables could not be included in any cluster at all). Moreover, a given clustering algorithm can produce disjoint or overlapping clusters.

In this work we focus on data clustering (exhaustive and disjoint) utilizing pairwise Euclidean distance between points in a m dimensional vectorial space.

Specifically, the objective is to partition a dataset $D = \{d_i | i = 1, \cdots, n\}$ with n records into a set of $k = |C|$ clusters according the following constraints: Every $d \in D$ has to be assigned to a cluster $C_i \in C$ such that $\forall_{C_i \in C} C_i \neq \emptyset$ and $\forall_{C_i, C_j \in C} C_i \cap C_j = \emptyset$. In this paper, we compare *AutoCPB* with the *K-means* [1] algorithm due to its popularity and robustness.

Several machine learning techniques have been applied to solve the problem of clustering. For example in [2], a neural network clusters data using entropy estimates. In [3], a genetic algorithm is combined with Nelder-Mead simplex search in order to produce a hybrid clustering algorithm. On the other hand, self organizing maps [4] have also been used for partitioning a dataset without specifying the number of clusters. However, much effort has not been dedicated to study evolutionary collaborative scheme for clustering. We propose an extension of the elemental particle swarm clustering algorithm for clustering.

Multiagent systems are used for simulating complex environments. In this paper, we propose a swarm intelligence clustering algorithm which takes advantage of simplistic communication and evolutionary methods in agents. Particle swarm optimization [5] is a class of evolutionary algorithms which aims to find a solution to a given optimization problem. Bacterial foraging algorithms [6] are a new paradigm in searching based on behavior of biological systems. In this paper we extend the advantages of social influence in PSO with the influence of bacterial foraging behavior.

The rest of this paper is organized as follows: Section 2 introduces background materials related to PSO and BFA. Section 3 describes our novel approach in detail. Section 4 depicts experimental results and some implementation details. Finally, Section 5 concludes our discussion and provides interesting remarks for future work.

2   Background

The fundamental idea of swarm intelligence algorithms [7] is that a set of individuals can cooperate in a decentralized manner increasing their productivity. Thus, the aim is to find mechanisms that can model complex systems, and represent them in a formal way [8].

2.1   Particle Swarm Optimization Exposed

Particle swarm optimization is a form of stochastic optimization based on swarms of simplistic, social agents [5]. Primary algorithms of particle swarm optimization perform search over a m dimensional space U by using a set of agents. In this lattice, an agent (particle) i occupy a position $x_i(t) = \{x_{i,j}(t) | j = 1, \cdots, m\}$ and has a velocity $v_i(t) = \{v_{i,j}(t) | j = 1, \cdots, m\}$ in an instant t, with a 1:1 correspondence (both $x_i(t)$ and $v_i(t)$ contain a set of components $\{j = 1, \cdots, m\}$ mapped to coordinates in U).

A simple PSO algorithm [7] works as follows: In the initialization phase, every agent takes positions around $x(0) = x_{min} + r(x_{max} - x_{min})$ and the velocities are set to 0 (x_{min}

and x_{max} are the minimal and maximal magnitudes in U and r is a real number between 0 and 1). Secondly, the algorithm enters in the search phase. The search phase consists of the following steps: Best cognitive/global position updating and velocity/position updating.

In the cognitive updating step every agent sets a value for the current (local) best position $y_i(t)$ that it has directly observed. The agent's local optima $y_i(t+1)$ is updated according to equation 1:

$$y_i(t+1) = \begin{cases} y_i(t) & \text{if } f(x_i(t+1)) \geq f(y_i(t)) \\ x_i(t+1) & \text{otherwise,} \end{cases} \tag{1}$$

whereas $f(x_i(t))$ is a fitness function which evaluates the goodness of a solution based on position $x_i(t)$.

In contrast, the global updating step sets in each iteration the best possible position Y_i observed by any agent. Equation 2 depicts the selection of the best global position.

$$Y_i(t+1) = \begin{cases} Y_i(t) & \text{if } f(y_i(t)) \geq f(Y_i(t)), \forall y_i(t) \\ y_i(t) & \text{if } \exists y_i(t)|f(y_i(t)) < f(Y_i(t)), \end{cases} \tag{2}$$

The velocity/position updating step selects new values for the position $x_i(t+1)$ and velocity $v_i(t+1)$ using equations 3 and 4 respectively:

$$x_i(t+1) = x_i(t) + v_i(t+1), \tag{3}$$

$$v_i(t+1) = v_i(t) + \underbrace{c_1 r_1 (y_i(t) - x_i(t))}_{\text{Cognitive component}} + \underbrace{c_2 r_2 (Y_i(t) - x_i(t))}_{\text{Global component}}, \tag{4}$$

where the c_1 and c_2 parameters are used to guide the search between local (cognitive component) and social (global component) observations. $r_1, r_2 \in [0,1]$ are random parameters which introduce a stochastic weight in the search. Finally, the algorithm stops until some convergence point is reached.

A suggested convergence test is proposed in [9]; it consists in testing whether $(f(Y_i(t)) - f(Y_i(t-1)))/f(Y_i(t))$ is smaller than a small constant ε for a given number of iterations. It is important to note that each particle in this multiagent system shares its knowledge (global best position) with all other particles by means of a neighborhood topology.

Several topologies have been proposed [7] (i.e. star, ring, clusters, Von Neumann, etc.). The difference between neighborhoods lies in how fast or slow (depending on connectivity) knowledge propagates through the swarm.

Further improvements have been proposed for the basic PSO algorithm [7,9]: Velocity clamping, inertia weight and constriction coefficient. All PSO-based algorithms in this study were implemented using the constriction coefficient (the most robust mechanism).

Velocity Clamping. A weakness of basic particle swarm optimization algorithms is that the velocity rapidly increases to unsuitable values. High velocity results in large position updates (in this case the particles may even leave the boundaries of the search space). This is specially true for particles occupying outlier positions. Velocity clamping is a simple restriction that imposes a maximal velocity. Equation 5 depicts this rule.

$$v_{i,j}(t+1) = \begin{cases} v'_i(t+1) & \text{if } v'_i(t+1) < V_{max} \\ V_{max} & \text{if } v'_i(t+1) \geq V_{max} \end{cases} \tag{5}$$

Obviously, large values of V_{max} facilitate exploration (small ones favor exploitation). V_{max} is frequently a fraction δ of the domain space for each dimension j in the search space U. Equation 6 presents this adjustment.

$$V_{max_j} = \delta(x_{max_j} - x_{min_j}), \tag{6}$$

whereas x_{max_j} and x_{min_j} represent the maximal and minimal magnitudes respectively.

Inertia Weight. The inertia weight is a constraint which aims to control the trade off between exploration and exploitation in a more direct fashion. Equation 7 introduces the inertia weight w that affects velocity updating.

$$v_i(t+1) = wv_i(t) + c_1r_1(y_i(t) - x_i(t)) + c_2r_2(Y_i(t) - x_i(t)) \tag{7}$$

$w \geq 1$ produces velocity increments over time (exploration), $w < 1$ decreases velocities over time (favoring exploitation). According to [7], the values $w = 0.7298$, $c_1 = c_2 = 1.49618$ have proved good results empirically, although in many cases they are problem dependent.

Constriction Coefficient. A similar method to inertia weight was proposed in [10], it is denominated the constriction coefficient. Equation 8 defines the new velocity updating mechanism.

$$v_i(t+1) = \chi(v_i(t) + \phi_1(y_i(t) - x_i(t)) + \phi_2(Y_i(t) - x_i(t))), \tag{8}$$

where $\phi = \phi_1 + \phi_2$, $\phi_1 = c_1r_1$, $\phi_2 = c_2r_2$ and

$$\chi = \frac{2k}{|2 - \phi - \sqrt{\phi^2 - 4\phi}|} \tag{9}$$

It is important to notice that $\phi \geq 4$ and $k \in [0,1]$ are necessary constraints to set χ in the range [0,1].

2.2   Bacterial Foraging Algorithms

Bacterial foraging algorithms are a new class of stochastic global search techniques [6]. Such algorithms emulate the foraging behavior of bacteria while situated in some nutrient substance. During foraging, a bacterium can exhibit two different actions: Tumbling or swimming.

The tumble action modifies the orientation of the bacterium. During swimming (chemotactic step) the bacterium will move in its current direction. After tumbling, the bacterium checks if it can find nutrients in its current direction, and if it can, then it will swim for a finite number of steps in that direction. After the bacterium has collected a given amount of nutrient, it will divide in two. The environment can also act in the bacteria population by eliminating or dispersing them.

A bacterial foraging algorithm can be defined as follows. Given a m dimensional search space U, each bacterium i has a position $x_i(t) = (x_{i,j}(t)|j = 1,\cdots,m)$ with m components at time t. It also has a chemotactic step size $C(i) > 0$ which influences the bacterium's step length. Initially the Bacteria is situated in several points in U. Then, each bacterium generates a random tumbling vector b_i consisting of m components. Following, the agent will enter into the swimming phase. Therefore, it will update its position one step at a time according to equation 10.

$$x_i(t+1) = x_i(t) + C(i) * b_i \qquad (10)$$

swimming will continue for a number N_a of iterations iff $f(x_i(t+1)) < f(x_i(t))$ holds. Once that all agents finished tumbling and swimming, they reproduce. New generations are created only with the half of the *healthiest* agents. *Healthy* agents can be interpreted as the ones which performed the smallest number of chemotactic steps. Finally, an elimination/dispersal step deletes and reallocates a percentage of agents at random. The algorithm will run for a fixed number of iterations.

A simple improvement for tumbling/swimming is to make bacteria attract each other in some area; and then repeal each other as they consume nearby nutrients. The idea is to calculate the cell to cell fitness, and add it to the fitness position for each bacterium.

3   PSO/BFA Multiagent Clustering

Previously, we presented the theoretical foundations necessary in which the novel *AutoCPB* relies on. In this section we introduce in detail our multiagent system for clustering.

One of the main problems related with PSO algorithms is that they have a tendency to fall into suboptimal solutions, because of the lack of diversity in the swarm. One of our ambitions is to find a mechanism to improve the diversity.

An improvement of the original PSO algorithm [11] is to use either a memetic approach, or hybridizing it with another swarm intelligence method [12]. Given that *AutoCPB* is a hybrid heuristic algorithm, we opted to design preliminary prototype-algorithms. We start introducing a simplistic swarm-based clustering method and then we gradually present more elaborated developments.

3.1   PSO Clustering

We implemented a simple clustering algorithm denominated *ClusterP*, which was proposed in [11]. Such method combines PSO and *K-means* for grouping data.

The dataset $D = \{d_{1,m}, d_{2,m}, \cdots, d_{n,m}\}$ defines the number of components m and instances n in the search space U. Thus, each datum $d_{i,j} \in D$ can be seen as a point in

U. Since the aim is to find a set of k desired clusters C containing every element in D; it is easy to visualize that the main problem is to find the set of $O = \{o_1, o_2, \cdots, o_k\}$ centroids that minimize the fitness function (Euclidean distance) with respect to each point in D. In *clusterP*, every agent $p_l \in P$ represent a solution with the set of position $O_l = \{o_{l,j} | j = 1, \cdots, k\}$. Algorithm 3.1 shows the *ClusterP* technique.

Algorithm 3.1. The *ClusterP* Algorithm.

Input: Data $D = \{d_1, d_2, \cdots, d_n\}, k$.
Output: Clusters $C = \{C_1, C_2, \cdots, C_k\}$.

1: Initialize swarm P (distribute O).
2: **REPEAT**:
3:     **FOR** $i = 1$ to n:
4:         **FOR** $j = 1$ to k:
5:             Calculate distances $dist(d_i, o_j)$.
6:         **END FOR**
7:         $C_h \Leftarrow d_i$ iff $argmin_{o_h \in O}(dist(d_i, o_h))$.
8:     **END FOR**
9:     **FOR** $l = 1$ to $|P|$:
10:         **Update** cognitive positions for p_l using equation 1.
11:         **Update** global position for P using equation 2.
12:     **END FOR**
13:     **FOR** $l = 1$ to $|P|$:
14:         **Update** velocities for p_l using equation 4.
15:         **Update** positions for p_l using equation 3.
16:     **END FOR**
17: **UNTIL** stopping condition holds.

ClusterP works as follows: It receives the data D and an integer k. First, it collocates every agent p in the environment, each containing a set of centroids O (line 1). Then, it assigns every record d_i to a cluster C_h iff the distance with its centroid $dist(d_i, o_h)$ is minimal (lines 3-8). Following, it obtains the cognitive and global position (lines 9-12), and updates every of its centroids velocities and positions (lines 13-16) as explained in Section 2.1. The algorithm stops after a fixed number of iterations or if no significant progress is made according to the fitness function.

3.2   Automatic PSO Clustering

ClusterP was extended in [13] in order to find the optimal number of clusters automatically. The new method is called *AutoCP*. It starts with a number k' of clusters ($k' \leq n$), and it deletes the *inconsistent* clusters.

We describe the form to detect inconsistent clusters as follows: First, for each cluster C_j we calculate its weight $W_j = \sum_{q=1}^{|C_j|} dist(o_j, d_q)$ as the sum of the distances from the center o_j to its points q_j where $q_j \neq o_j$. Immediately, all weights W from the clusters are sorted. Then, we normalize all weights W_j with respect to the cluster with lowest value W_s, such that the set of local thresholds become $th = \{th_j | j = 1, \cdots, k', th_j = W_s/W_j, W_s = argmin_{w \in W}(w)\}$. Finally, a cluster C_j is declared as *inconsistent* iff its threshold t_j is lower than the global threshold T, whereas T is in the range of $[0,1]$. Algorithm 3.2 introduces *AutoCP*.

Algorithm 3.2. The *AutoCP* Algorithm.

Input: Data $D = \{d_1, d_2, \cdots, d_n\}$.
Output: Clusters $C = \{C_1, C_2, \cdots, C_{k'}\}$.

1: Initialize swarm P (distribute O).
2: **REPEAT**:
3:     Assign D to clusters as in algorithm 3.1.
4:     **Update** cognitive/global positions as in algorithm 3.1.
5:     **Update** velocities/positions as in algorithm 3.1.
6:     **FOR** $j = 1$ to k':
7:         Calculate W_j for each cluster C_j.
8:     **END FOR**
9:     Obtain $W_s = argmin_{W_j \in W}(W_j)$.
10: **FOR** $j = 1$ to k':
11:     $th_j = W_s/W_j$.
12:     **IF** $(th_j < T)$
13:         Remove C_j and $k' = k' - 1$.
14:     **END IF**
15: **END FOR**
16: **UNTIL** stopping condition holds.

As previously mentioned, *AutoCP* simply adds a mechanism to *ClusterP* for automatic detection of clusters with no further modification (lines 6-15). At the end of each iteration, we select and discard inconsistent clusters.

3.3   AutoCB Clustering

In this section we present the method for automatic clustering using bacterial foraging algorithm *AutoCB*. It is an adapted version of the bacterial foraging algorithm for automatic clustering. Basically, it uses the *K-means* principle to add data to the closest centroids combined with a bacterial foraging search.

In algorithm 3.3, every agent observes a clustering solution represented by the positions of bacteria as centroids (notice that times are references to current $x_i(t)$ or posterior $x_i(t+1)$ positions in space. x_i in fact represent the tentative position for centroid o_i). In order to keep this algorithm as simple as possible, we decided not to add any complex initialization method. We simply exchanged the swimming mechanism to be executed firstly, followed by tumbling. In this way we use the fitness function to set up

Algorithm 3.3. The *AutoCB* Algorithm.

Input: Data $D = \{d_1, d_2, \cdots, d_n\}$.
Output: Clusters $C = \{C_1, C_2, \cdots, C_{k'}\}$.

1: Initialize B (distribute O randomly).
2: **FOR** Number of Elimination/Dispersal Steps
3:     **FOR** Number of Reproduction Steps
4:         **FOR** Number of Chemotactic Steps
5:             **FOR** Each Bacterium $i \in B$
6:                 **WHILE** $m < MaxSwimLength$
7:                     Delete inc. clusters as in algorithm 3.2.
8:                     Assign D to C as in algorithm 3.1.
9:                     **IF** $f(x_i(t+1), o_i) < f(x_i(t), o_i)$
10:                        $o_i = x_i(t+1)$.
11:                        $m = m+1$.
12:                     **END IF**
13:                     **ELSE**
14:                        $m = MaxSwimLength$.
15:                     **END ELSE**
16:                 **END WHILE**
17:                 $tumble(i)$, generate new $x_i(t+1)$.
18:             **END FOR**
19:         **END FOR**
20:         $reproduce(B)$.
21:     **END FOR**
22:     $eliminate(B), disperse(B)$.
23: **END FOR**

bacteria in good spots since the beginning. Initially, the algorithm distributes the positions for all centroids of every agent in B at random (line 1). For every bacterium i, the chemotactic part of the algorithm starts (lines 4-19): The agents will update their centroids positions for a maximal number of swims *MaxSwimLength* (lines 6-16). Firstly, the algorithm deletes inconsistent clusters as mentioned in lines 6-15 of algorithm 3.2 (line 7). Then, we assign each datapoint $d \in D$ to a cluster $C_j \in C$ according to the lines 3-8 of algorithm 3.1 (line 8).

Every agent observes the cells $x_i(t+1)$ in front of its centroids o_i. Then, they swim and update their centroids positions from o_i to $x_i(t+1)$ iff the Euclidean distance $f(x_i(t+1), o_i)$ to the new point is smaller than the current one $f(x_i(t), o_i)$ (lines 9-12). Swimming might immediately stop according to the control variable m in line 14; if the fitness of the tentative cell $f(x_i(t+1), o_i)$ is greater or equal to the one of the current position $f(x_i(t), o_i)$.

Once agent i has finished swimming, it will *tumble* (line 17). Thus, i will choose a tentative cell in a new direction $x_i(t+1)$ for each of its centroids $o_i = x_i(t)$. $x_i(t+1)$

is selected at random. Finally, the previous chemotactic process is encapsulated in a reproductive phase (line 3), and in a combined elimination/dispersion phase(line 2). The reproduction process deletes half of the agents having the smallest number of chemotactic steps. The elimination method destroys again a percentage α of agents at random. Dispersion reallocates a small percentage β of the remaining agents in U at random.

3.4   The AutoCPB Algorithm

The previous *AutoCB* algorithm is strictly based on the theory of bacterial foraging algorithms. However, we believe that we can improve its performance by sharpening the swimming and tumbling phase using a PSO-based algorithm.

In this section we describe in detail the hybrid algorithm for clustering based on PSO and BFA denominated *AutoCPB*. The algorithm follows the same structure of the AutoCB algorithm. Nevertheless, each bacterium agent i has assigned a position and a velocity to its centroids as previously seen in algorithm 3.1. The tumble is now guided by the swarm local and social beliefs.

A tentative cell $x_i(t+1)$ is deterministically chosen according to the best neighboring cell $y_i(t+1)$ and the global best position $Y_i(t+1)$. Swimming is still performed for a given number of iterations but now the agent's steps vary in dimension according to its velocity. *AutoCPB* contains a bacterial foraging skeleton and a particle swarm optimizator. In this algorithm, we simply replace tumbling in line 17 from algorithm 3.3 by a more elaborated PSO-based foraging mechanism (lines 18-24). Logically, swimming in lines 9-12 of algorithm 3.3 is also modified so we take advantage of the guided tumbling.

Specifically, the *AutoCPB* clustering algorithm works as follows: It collocates every centroid o_i for agent i in U at random $(x_i(t) = o_i)$. Then, the modified tumbling/ swimming version is executed for all agents for a number of swims *MaxSwimLength* (lines 4-25): Inconsistent clusters are removed (line 7) as seen in algorithm 3.2. All datapoints $d \in D$ are assigned to the remaining clusters $C = \{C_j | j = 1, \cdots, k'\}$ (line 8). Then, all centroid positions $o_i = x_i(t)$ for agent i will be updated with the new position $x_i(t+1)$ iff $f(x_i(t+1)) < f(x_i(t))$ holds (lines 9-10). Otherwise, swimming stops in line 16. In this part of the algorithm, we swim by updating the value of $x_i(t+1)$ according equation 3 (line 11). We also record the best cognitive/local position $y_i(t+1)$ according to equation 1 for further tumbling computation (line 12). The final step of swimming is the update of the best global/social position $Y_i(t+1)$ (line 18).

In lines 21-24, PSO-based tumbling is executed. At this point, every centroid (with a velocity $v_i(t)$ and a position $x_i(t)$) starts observing its neighboring cells. It finally updates its own velocity to $v_i(t+1)$ and tentative position $x_i(t+1)$ by using equations 3 and 4 respectively. Notice that agents do not modify its current position during tumbling but only during the swimming phase (lines 9-14). Finally, once that we have clustered U into C, *AutoCPB* will reproduce (line 26), eliminate and disperse (line 28) agents in the same manner *AutoCB* does.

In the next part of this paper we comprehensively tested every approach obtaining promising results.

Algorithm 3.4. The *AutoCPB* Algorithm.

Input: Data $D = \{d_1, d_2, \cdots, d_n\}$.
Output: Clusters $C = \{C_1, C_2, \cdots, C_{k'}\}$.

1: Initialize B (distribute O randomly).
2: FOR Number of Elimination/Dispersal Steps
3:   **FOR** Number of Reproduction Steps
4:   **FOR** Number of Chemotactic Steps
5:   **FOR** Each agent $i \in B$
6:   **WHILE** $m < MaxSwimLength$
7:   Delete inconsistent clusters as in algorithm 3.2.
8:   Assign D to C as in algorithm 3.1.
9:   **IF** $f(x_i(t+1)) < f(x_i(t))$
10: $o_i = x_i(t+1)$.
11:   **Update** $x_i(t+1)$ using equation 3.
12:   **Update** $y_i(t+1)$ using equation 1.
13: $m = m + 1$.
14:   **END IF**
15:   **ELSE**
16: $m = MaxSwimLength$.
17:   **END ELSE**
18:   **Update** $Y_i(t+1)$ using equation 2.
19:   **END WHILE**
20:   **END FOR**
21:   **FOR** Each agent $i \in B$
22:   **Update** $v_i(t+1)$ using equation 4.
23:   **Update** $x_i(t+1)$ using equation 3.
24:   **END FOR**
25:   **END FOR**
26:   *reproduce*(B).
27:   **END FOR**
28:   *eliminate*$(B), disperse(B)$.
29: END FOR

4   Experimental Results

We proceed to analyze the performance of the aforementioned algorithms by testing them over well known benchmark datasets. Before we introduce the results, we describe the datasets, parameter settings and performed tests used for experimentation.

The algorithms were tested on nine datasets (D). Two domains were generated at random: Artificial1 (A1) and Artificial2 (A2). The rest of the datasets were taken from the UCI repository [14]: Iris (Ir), Wine (Wi), Pima (Pi), Haberman (Ha), BreastCancer (BC), Glass (Gl), and Yeast (Ye). Table 1 introduces the characteristics for each dataset in terms of its number of classes and dimensions.

Table 1. The datasets used for experimentation

D	Classes	Dimensions
A1	2	2
A2	3	3
Ir	3	4
Wi	3	13
Pi	2	8
Ha	2	3
BC	2	30
Gl	6	9
Ye	10	8

4.1   Environmental Settings

Each experiment is based on fifty consecutive runs of the same algorithm. For the *K-means* algorithm we set a maximal of 500 iterations (it terminates if no improvement is made). In the other algorithms, the assignation of points to clusters consists of a single iteration.

For *ClusterP* and *AutoCP* the stopping condition is the one described in Section 2.1. For *AutoCB* and *AutoCPB* we used 10 elimination/dispersal iterations, 20 reproduction steps and 20 chemotactic steps. In every method we used 30 agents. Every dataset has a predefined class. Therefore, we set k in *ClusterP* as the number of classes (in fact, we found that K-means obtains the best results in this fashion). For the multiagent-based automatic clustering algorithms we used 30 agents and an upper limit k' of 10 initial clusters.

4.2   Cluster Validation

According to [15] cluster validation can be divided into external, internal and relative validation. When using an external measure, we compare the solution with some optimal solution known a priori. Internal validation deals with judging the solution based on the intra-distance of a cluster. Relative measures compare the experimental clustering solution against another set of clusters previously found.

We used two validation measures, namely the inter cluster distance measure (ID) and the quantization error function (QEF). Both metrics have previously utilized to test cluster reliability [11]. The inter cluster distance calculates the average distances between the centroids and all their points in the cluster. It is calculated according to equation 11.

$$ID_C = \sum_{\forall o_i, o_j \in CEN, i \neq j} dist(o_i, o_j)/|C|, \tag{11}$$

where o_i and o_j are centroids, CEN is the set of all centroids, $|C|$ is the total number of clusters C and $dist()$ is the Euclidean distance. In essence, we calculate the average Euclidean distance between all pairs of centroids. The QEF is a global distance measure that evaluates the average distance from all datapoints to centroids in every cluster. Equation 12 present the QEF metric.

$$QEF_C = \Big(\sum_{j=1,k} \Big(\sum_{\forall d_i \in C_j} dist(d_i, o_j)/|C_j| \Big) \Big)/k, \tag{12}$$

whereas k is the number of clusters, d_i is a datapoint contained in cluster C_j. All other variables are defined as in equation 11. The previous measures can be used to express the quality of a solution. Thus, we proceed to establish a discussion of the performance of the algorithms.

4.3   Benchmark Testing

Table 2 includes the average results for the inter cluster distance, quantization error function, number of clusters and the elapsed times (in milliseconds). Numbers in bold represent the best result for each dataset.

Even though, it can be reasoned as subjective to evaluate different clustering methods, we are confident that our test reflect some truth regarding the quality of the clustering. In fact, we believe that the ID and QEF tests express an acceptable comparison between clusterings for this investigation. Thus, every algorithm makes use of the *K-means* oriented mechanism for assigning points to clusters.

Specifically, every clustering method uses the **same** objective function. We add a datapoint to a cluster whose cluster centroid is the closest. The only difference between algorithms is their search mechanism, not contemplating any other characteristic for grouping. For all cases, we appreciate that *K-means* is the worst performing algorithm in terms of quality and the best in running time. Thus, every swarm based algorithm adds a refined search to *K-means*. However, in many cases we are more concerned in the quality of a solution. We can also improve the implementation of the algorithms in order to decrease the elapsed times.

We can conclude that with respect to the QEF metric; *AutoCP* and *AutoCPB* are the best algorithms, finding minimal values. *AutoCP* having a minimal QEF in A1, A2, Ir, Gl and Ye. *AutoCPB* is the most competitive in Ha, Pi, Wi, BC and Ye. However, both methods produce similar results. In every dataset any of them can be the first and the second most competitive method. This behavior is logical, and it is a common example of how the random element introduces some noise to a search technique. *AutoCPB* performs random steps (reproduction, elimination and dispersion).

For the ID metric we observe an almost identical trend. *AutoCB* outperforms all other methods in some cases (A1, A2). Indeed, *AutoCB* is the most randomized algorithm, it performs a nearest neighbor search with poor guidance but with a dynamical evolutionary mechanism. In every case, we can see that *ClusterP* is suboptimal and falls into local optima. The later is one of the reasons why the evolutionary method in the form of automatic clustering and bacterial foraging show a promising enhancement.

A common discussion in data mining is related with the selection of the *appropriate* number of clusters k for a given domain. A common assumption is that we should apply a given clustering technique depending on the particular problem/domain to analyze. However, with the advents in information technology and the analytical capabilities of

Table 2. Full clustering results. Swarm-based enhancements improve the performance of *K-means*

D	Method	QEF	ID	$|C|$	E. time
A1	*K-means*	11.64 +/-0.10	28.4 +/-0	2	**14.06 +/-5.75**
A1	*ClusterP*	11.99 +/-0.42	29.95 +/-2.03	2	1394.68 +/-18.7
A1	*AutoCP*	**4.91 +/-0.57**	4.13 +/-2.46	5.6 +1.01	2105.08 +/-427.96
A1	*AutoCB*	7.75 +/-1.18	**3.36 +/-2.31**	4.48 +/-1.27	8439.36 +/-423.15
A1	*AutoCPB*	6.2 +/-0.7	3.73 +/-2.51	4.8 +/-0.76	835.3 +/-74.6
A2	*K-means*	44.95 +/-33.56	28.89 +/-11.26	3	**95.3 +/-7.36**
A2	*ClusterP*	29.92 +/-2.27	35.62 +/-7.2	3	6914.04 +/-59.66
A2	*AutoCP*	**24.36 +/-1.13**	8.09 +/-3.46	4.9 +/-1.06	9446.92 +/-1376.23
A2	*AutoCB*	30.40 +/-3.07	**7.14 +/-3.87**	4.36 +/-1.03	11914.98 +/-269.82
A2	*AutoCPB*	24.45 +/-2.11	7.73 +/-3.94	4.22 +/-0.93	3658.74 +/-298.91
Ha	*K-means*	14.59 +/-6.69	8.85 +/-0.13	2	**135.94 +/-7.25**
Ha	*ClusterP*	10.23 +/-0.32	6.83 +/-1.96	2	13655.02 +/-171.18
Ha	*AutoCP*	9.18 +/-0.65	1.39 +/-0.84	3.26 +/-0.83	18916.88 +/-3962.37
Ha	*AutoCB*	12.41 +/-1.73	1.57 +/-0.82	2.98 +/-0.98	12055.92 +/-272.07
Ha	*AutoCPB*	**8.87 +/-0.76**	**1.2 +/-0.7**	3.14 +/-0.64	6930 +/-366.89
Pi	*K-means*	132.86 +/-46.72	111.77 +0	2	**742.2 +/-15.54**
Pi	*ClusterP*	73.85 +/-3.363	35.25 +/-11.43	2	72757.18 +/-1564.94
Pi	*AutoCP*	72.35 +/-3.29	11.43 +/-12.17	3.02 +/-1.12	107948.8 +/-20853.2
Pi	*AutoCB*	96.33 +/-12.68	12.06 +/-11.92	3.16 +/-0.89	33195.9 +/-1179.08
Pi	*AutoCPB*	**65.43 +/-6.73**	**6.65 +/-5.6**	2.96 +/-0.53	37445.02 +/-4117.43
Ir	*K-means*	1.58 +/-0.58	1.13 +/-0.49	3	**117.5 +/-7.89**
Ir	*ClusterP*	0.98 +/-0.14	1.19 +/-0.33	3	8390.32 +/-91.91
Ir	*AutoCP*	**0.78 +/-0.06**	0.26 +/-0.15	3.8 +/-0.81	11591.68 +/-1577.85
Ir	*AutoCB*	1.13 +/-0.17	0.25 +/-0.16	3.46 +/-1.16	14862.18 +/-297.83
Ir	*AutoCPB*	0.84 +/-0.09	**0.24 +/-0.13**	3.62 +/-0.81	4303.76 +/-324.01
Wi	*K-means*	216.67 +/-52.69	162.28 +/-63.28	3	**403.76 +/-13.2**
Wi	*ClusterP*	131.5 +/-8.60	68.74 +/-43.93	3	26301.88 +/-546.39
Wi	*AutoCP*	99.49 +/-7.02	24.79 +/-18.72	5.48 +/-1.05	37724.1 +/-3461.18
Wi	*AutoCB*	141.87 +/-17.68	32.1 +/-22.11	4.54 +/-1.11	42410.36 +/-1111.42
Wi	*AutoCPB*	**97.87 +/-10.51**	**23.29 +/-21.49**	4.88 +/-1.1	13421.86 +/-799.42
BC	*K-means*	563.32 +/-142.04	665.67 +/-0	2	**1949.08 +/-215.4**
BC	*ClusterP*	314.87 +/-11.53	141.46 +/-51.5	2	174520.6 +/-3574.57
BC	*AutoCP*	196.4 +/-14.18	44.3 +/-41.07	5.3 +/-1.04	259376.08 +/-34956.1
BC	*AutoCB*	292.89 +/-53.18	59.86 +/-50.86	4.32 +/-1.28	108388.1 +/-2871.43
BC	*AutoCPB*	**195.01 +/-22.4**	**34.44 +/-35.2**	4.62 +/-0.83	92403.2 +/-8281.34
Gl	*K-means*	9.23 +/-7.21	0.63 +/-0.29	6	**698.74 +/-41.74**
Gl	*ClusterP*	3.31 +/-0.42	0.39 +/-0.22	6	25520.92 +/-280.95
Gl	*AutoCP*	**1.23 +/-0.25**	0.29 +/-0.24	2.28 +/-0.5	45504.3 +/-24056.07
Gl	*AutoCB*	2.31 +/-0.4	0.24 +/-0.16	2.94 +/-1.27	29877.22 +/-651.5
Gl	*AutoCPB*	1.62 +/-0.31	**0.21 +/-0.2**	2.2 +/-0.45	11431.86 +/-1466.4
Ye	*K-means*	1.46 +/-1.09	0.04 +/-0.02	10	**6815.02 +/-62.67**
Ye	*ClusterP*	0.8 +/-0.12	**0.03 +/-0.01**	10	175838.7 +/-1397.52
Ye	*AutoCP*	**0.26 +/-0.02**	**0.03 +/-0.01**	2.12 +/-0.33	211558.7 +/-46305.2
Ye	*AutoCB*	0.34 +/-0.04	0.04 +/-0.02	2.42 +/-0.61	35482.18 +/-558.49
Ye	*AutoCPB*	**0.26 +/-0.02**	**0.03 +/-0.02**	2.22 +/-0.42	69598.58 +/-3044,74

the different branches of science we find enormous amounts of data from phenomena we can not truly understand (i.e. complex gene expression data, astronomical data). The later, is a true motivation for the development of automatic clustering algorithms.

From our results we can argue that in most domains, as the number of clusters increase, the number of data points diminish. However, this can not be true in the cases where the optimal clustering comprises a very segregated and marginal domain.

We can observe in Table 1 that a higher number of clusters (A1, A2, Ha, Pi, Ir, Wi and BC) tends to give a lower result in both ID and QEF. This might be because our tests do not express what we are really looking for, or perhaps, that we have discovered a new feature in the dataset. Thus, we can modify our assumptions about the optimal number of clusters in a dataset. Actually, since we use the same cluster assignment method in all techniques, we can conclude that we have also found the optimal number of clusters k given our objective function.

5   Final Discussion

In this work we described two swarm intelligence techniques, namely particle swarm optimization and bacterial foraging. We described in detail several algorithms based on these paradigms. The new algorithms improve the basic *ClusterP* method which in fact is based on *K-means*.

The essence of our work is that we managed to combine two major paradigms PSO and BFA in order to create a robust clustering algorithm. The resulting hybrid method *AutoCPB* showed improvements during experimentation. The original *AutoCP* algorithm was unable to reach the best results on its own. Actually, *AutoCB* is also suboptimal in non-random datasets because of its swarm model, which does not perform global search. Therefore, we sustained with empirical results that by properly combining the two techniques result in a true enhancement for clustering. Thus, we take the best feature of each algorithm (the *smart* foraging in PSO and the evolutionary rules from BFA).

The testing was possible due to our implemented framework, which enabled us to test all algorithm in well known datasets. In general, we believe that the application of hybrid PSO/BFA mechanisms is promising. We expect to improve AutoCPB to a major extent.

Future work consists in identifying a rule to minimize local optima in AutoCPB. This algorithm can be applied to other domains such as attribute clustering. This stochastic/evolutionary method can also show features not only in data but between variables in datasets. A more specific analysis of parameter setting is also considered as a tentative improvement.

References

1. Han, F., Kamber, K.: Data mining: Concepts and techniques, pp. 334–395. Academic Press, San Francisco (2001)
2. Tazutov, A., Kurenkov, N.: Neural network data clustering on the basis of scale invariant entropy. In: International Joint Conference on Neural Networks, pp. 4912–4918 (2006)

3. Chandra, S., Murthy, J.: An Efficient Hybrid Algorithm for Data Clustering Using Improved Genetic Algorithm and Nelder Mead Simplex Search. In: Proc. of the Int. Conf. on Comp. Int. and Mult. Appl., vol. 1, pp. 498–510 (2007)
4. Goncalves, M., Andrade, M.: Data Clustering using Self-Organizing Maps segmented by Mathematic Morphology and Simplified Cluster Validity Indexes: an application in remotely sensed images. In: International Joint Conference on Neural Networks, pp. 4421–4428 (2006)
5. Kennedy, J., Ebenhart, R.: Particle Swarm Optimization. In: Proceedings of the 1995 IEEE Int. Conf. on Neural Networks, pp. 1942–1948 (1995)
6. Passino, K.: Biomimicry of bacterial foraging for distributed optimization and control. Control Systems Magazine, 52–67 (2002)
7. Engelbrecht, A.: Fundamentals of Computational Swarm Intelligence. Wiley, Chichester (2005)
8. Heylighen, F., Joslyn, C.: Cybernetics and second-order cybernetics. Encyclopedia of Physical Science and Technology. Academic Press, New York (2001)
9. Bergh, F.: An analysis of particle swarm optimizers. PhD Thesis, University of Pretoria, South Africa (2002)
10. Clerc, M., Kennedy, J.: The particle swarm: Explosion, stability and convergence. IEEE Transactions on Evolutionary Computation 6, 58–73 (2002)
11. van der Merwe, D., Engelbrecht, A.: Data Clustering using particle swarm optimization. In: The 2003 congress on Evolutionary Computation, vol. 1, pp. 215–220 (2003)
12. Biswas, A., Dasgupta, S.: Synergy of PSO and bacterial foraging optimization: A comprehensive study on. Advances in Soft Computing Series, pp. 255–263. Springer, Heidelberg (2007)
13. Abraham, A., Roy, S.: Swarm intelligence algorithms for data clustering. Chp. 12, 279–313 (2008)
14. Asuncion, A., Newman, D.: UCI Machine Learning Repository. University of California, Irvine, School of Information and Computer Sciences (2007), http://www.ics.uci.edu/~mlearn/MLRepository.html
15. Theodoris, S., Koutroumbas, K.: Pattern Recognition. Academic Press, London (2006)

A Self-Organized Multiagent System for Intrusion Detection

Esteban J. Palomo, Enrique Domínguez, Rafael M. Luque, and Jose Muñoz

Department of Computer Science
E.T.S.I. Informatica, University of Malaga
Campus Teatinos s/n, 29071 – Malaga, Spain
{ejpalomo,enriqued,rmluque,munozp}@lcc.uma.es

Abstract. This paper describes a multiagent system with capabilities to analyze and discover knowledge gathered from distributed agents. These enhanced capabilities are obtained through a dynamic self-organizing map and a multiagent communication system. The central administrator agent dynamically obtains information about the attacks or intrusions from the distributed agents and maintains a knowledge pool using a proposed growing self-organizing map. The approach integrates traditional mathematical and data mining techniques with a multiagent system. The proposed system is used to build an intrusion detection system (IDS) as a network security application. Finally, experimental results are presented to confirm the good performance of the proposed system.

Keywords: Multiagent system, data mining, growing self-organization, network security.

1 Introduction

Multiagent systems are a very active field to handle the dynamism in the environment due to its modularity and its flexibility. These characteristics are essentials in complex, large and distributed domains. Network security applications are provided with an enormous amount of data from which the systems must be capable to detect anomalies, intrusions or attacks. Therefore, modularity and flexibility offered by multiagent systems can be applied to network security to obtain better results.

The proposed system carries out a reasoning of the domain using local knowledge. Several multiagent systems for knowledge discovery have recently been proposed [2, 10, 12]. Multiple agents acting on different environments extract a local knowledge and communicate this information back to the central administrator agent to form a knowledge pool about the domain. In this sense, distributed agents are specialized at detecting intrusions or attacks while the central administrator agent maintains a knowledge base of the different anomalies together with the normal behavior of the system. The data in the knowledge base are clustered and analyzed to identify similarities and variations between the anomalies and normal behaviors. A proposed growing SOM model is used for clustering data in the knowledge base.

The self-organizing map (SOM) has been widely used as a unsupervised learning method for clustering high-dimensional input data and mapping these data into a two-dimensional representation space [4]. This mapping preserves the topology in terms of

L. Cao et al. (Eds.): ADMI 2009, LNCS 5680, pp. 84–94, 2009.

distances in the representation space and provides an easy visualization of clusters on the map. One limitation of this model is its static network architecture, where topology and number of nodes have to be established in advance. This task can be difficult because a prior study of the problem domain, especially when we have vectors of high dimensionality, is needed.

The problem of determining the number of nodes of a SOM has been addressed by different SOM models. Some of them are the incremental grid growing, growing grid, growing SOM, and the hypercubical SOM. These models have in common the addition of new nodes in the neighborhood of others, which are considered as bad representatives of data mapped into them. However, the generated topology is not as flexible as possible. The growing grid adds rows or columns in a 2-D grid, so the topology is always a rectangular grid. There are wide spectrum application domains where a rectangular grid is not the most suitable topology to represent the input data. In order to make easy the identification of relations among input data and mirror its inherent structure, the map should capture the own topology of the data as faithfully as possible. The hypercubical SOM grows into more than two dimensions, but forsakes visual understanding of data. The rest of the models have a more adaptive topology but are limited to a growth in 4 directions at the boundary of the map. Moreover, the inherent hierarchical structure of data is not represented for these models.

In this paper, a growing SOM with a more flexible adaptation of its topology to represent input data is presented. This new model splits the nodes with highest representation error, taking into account their more dissimilar neighbors. Furthermore, hierarchical relations among data are provided by classifying the prototype vectors of clusters. Therefore, a two-level hierarchy of input data is mapped in a 2-D representation space.

The remainder of this paper is organized as follows. Section 2 contains a detailed description of the proposed system architecture for network security. In Section 3, the new growing SOM model for clustering data in the knowledge base is presented. Experimental results from implementing an IDS with the proposed multiagent system are provided in Section 4. Finally, Section 5 concludes this paper.

2   Overview of the Proposed Multiagent System

Network security application have complex and distributed domains with a huge amount of data from which the attacks or anomalies must be detected. Therefore, modularity and flexibility offered by multiagent systems are suitable characteristics to obtain good results.

An IDS monitors the IP packets flowing over the network to capture intrusions or anomalies, where anomalies are considered deviations from the normal behavior. Therefore, the proposed system detects anomalies or attacks by classifying IP connections into normal or anomalies connections. Moreover, the proposed system also identifies the type of known attacks. Some anomaly detection systems using data mining techniques such as clustering, support vector machines (SVM) and neural network systems have been proposed [5, 6, 9]. The artificial neural networks provide many advantages in the detection of network intrusions [1]. However, IDSs based on self-organization usually show high false positive rates [11].

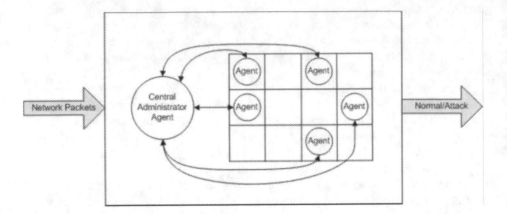

Fig. 1. IDS based on a self-organized multiagent system

The proposed multiagent architecture is composed by multiple distributed agents and a central administrator agent, as shown in Fig. 1. Distributed agents detect anomalies or attacks. The central administrator keeps a global knowledge that is accessible by the other agents for further information about the domain. Similarly, the central administrator agent can transfer the knowledge base to the distributed agents to enhance the intrusion detection capability of individual agents.

Knowledge base maintenance using some similarity measure is essential for detecting common patterns or behaviors and for generalizing to new attacks. Therefore, the central administrator agent uses a clustering technique based on the proposed growing SOM to represent the global knowledge. The novelty of the proposed architecture is mainly with the hybridization of mathematical techniques and the proposed clustering technique in a multiagent system to implement an IDS.

3 New Growing SOM Model

The new growing SOM is a novel neural network for clustering high-dimensional input data and mapping these data into a two-dimensional representation space, where this representation space adapts its topology as faithfully as possible to input data and hierarchical relations among data are represented as groups of clusters.

This model is initialized to a SOM of 2x2 nodes, where each node represents an agent. Then, the neural network is trained through the usual feature map self-organizing process, adapting the weight vectors of nodes w_i according to input data x as the following expression

$$w_i(t+1) = w_i(t) + \alpha(t)h_i(t)[x(t) - w_i(t)] \tag{1}$$

where α is the learning rate decreasing in time, h_i is a Gaussian neighborhood function that reduces its neighborhood kernel at each iteration and t denotes the current training iteration.

Once training ends, the splitting process of the nodes of the map begins. Here, the representation quality of each node is checked. The representation quality of a node is also called quantization error (qe), which is a measure of the similarity of data mapped onto each node, where the higher is the qe, the higher is the heterogeneity of the data cluster. The quantization error of the node i is defined as follows

$$qe_i = \sum_{x_j \in C_i} \|w_i - x_j\| \tag{2}$$

where C_i is the set of patterns mapped onto the node i, x_j is the jth input pattern from C_i, and w_i is the weight vector of the node i.

The first step of the training algorithm is to compute the initial quantization error qe_0 of the map. This quantization error measures the dissimilarity of all input data and it is used for the growth process of the SOM together with the τ parameter, following the splitting condition given in (3).

$$qe_i < \tau \cdot qe_0 \tag{3}$$

This way, the quantization error of a node i (qe_i) must be smaller than a fraction τ of the initial quantization error (qe_0). Otherwise, the node is considered an error node (e), indicating that their mapped data are heterogeneous enough to be represented for more than one node. Then, these nodes are split into two nodes, so just one node is added for each error node. Thus, the parameter τ controls the growth process of the map, so the smaller the parameter the bigger the map and vice versa.

Before an error node is split, we have to decide the location where the two nodes are placed. This location is determined by their two more similar neighbors. Here, the neighbors of a node are the closest nodes in the same row, column or diagonal, so a node can have up to 8 possible neighbors depending on its position at the map (see Fig. 2(a)). This way, the indexes of the two more similar neighbors of the error node (s_1 and s_2) are chosen as follows:

$$
\begin{aligned}
s_1 &= \arg\min_i(\|w_e - w_i\|), & w_i &\in \Lambda_e \\
s_2 &= \arg\min_i(\|w_e - w_i\|), & w_i &\in \Lambda_e - \{w_{s_1}\}
\end{aligned}
\tag{4}
$$

where w_e is the weight vector of the error node, w_i is the weight vector of the ith node, and Λ_e is the set of neighbor nodes of e. Note that in finding the second more similar neighbor, Λ_e is the set of neighbor nodes of e except the most similar neighbor s_1.

The new neurons are placed between the error node e and their most similar nodes s_1 and s_2. If between these nodes there is no space, a conflict is happened, and a row and/or a column is created to solve the conflict and place the new nodes. All the possible conflicts are shown in Fig 2.

Since more than one node can be divided after training, when conflicts are present we have to take into account the rest of dividing nodes in order to avoid add unnecessary rows or columns. An example of this splitting process is shown in Fig. 3.

The weight vectors of the new nodes are initialized as the mean of the weight vectors of the error node and the chosen neighbor. If two error nodes have as most similar neighbor the other error node, as A and B nodes in Fig. 3, the new weight vector is $2/3$

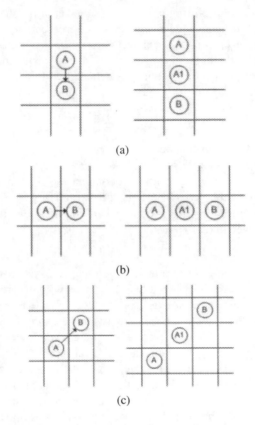

Fig. 2. Different splitting conflicts: (a) A row or (b) a column or (c) a row and a column are inserted between A and B, where A is the error node and B is one of its most similar neighbors

the weight vector of the error node and $1/3$ the weight vector of the neighbor. In both cases, the weight vector will be initialized as follows:

$$w_{new} = c_1 w_e + c_2 w_s, \qquad w_s \in \{s_1, s_2\} \tag{5}$$

where $c_1 = c_2 = 1/2$ if node s is not a error node, and $c_1 = 2/3$ and $c_2 = 1/3$ if node s is also a error node and one of their two most similar neighbors is e.

After the map has grown with the splitting process, the SOM is trained again and the entire process is repeated until the splitting condition (3) is satisfied for all nodes. Then, hierarchy relations must be represented in the representation space. For that reason, a clustering of clusters is performed at this stage. This clustering is performed as the usual self-organizing process, but considering weight vectors of nodes as input data. As consequence, groupings of clusters will be discovered providing information about a two-level hierarchy from input data. This is a bottom-up hierarchy discovering, where inherent hierarchical relations are represented at two levels.

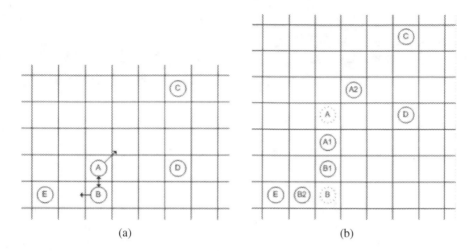

(a)   (b)

Fig. 3. An example of the splitting process with conflicts: (a) Shows a grid with 5 nodes, where A and B are the error nodes and the arrows indicate the direction of their two most similar neighbors (B and C for A, and A and E for B). Note that two row conflicts between A and B are present. (b) Shows the grid after the node splitting, where two rows has been added to solve the two row conflicts between A and B, and four new nodes have been created (A1, A2, B1 and B2). After splitting, A and B has been deleted.

4   Experimental Results

The proposed multiagent system has been proved by implementing an Intrusion Detection System (IDS) . The KDD Cup 1999 benchmark data set has been used. It was created by MIT Lincoln Laboratory and contains a wide variety of instrusions simulated in a military network environment. To make easy the implementation, the 10% KDD Cup 1999 benchmark training and testing data set has been selected, which contain 494,021 and 311,029 connection records, respectively. Each connection record is composed of 41 pre-processed features from monitoring TCP packets. In the training data set exists 22 attack types in addition to normal records, which fall into four main categories [3]: DoS (denial of service), Probe, R2L (Remote-to-Local) and U2R (User-to-Root). In the testing data set 15 new attack types are found.

Both data sets contain three qualitative features in addition to quantitative features: protocol type, service and status of the connection flag. These qualitative features have to be mapped to quantitative values in order to compare two vectors with the Euclidean distance. However, qualitative values cannot be directly mapped into quantitative values since it assigns an order among qualitative values. Therefore, each qualitative feature has been replaced with new binary features (dummy features), which represent the possible values of each qualitative feature. Thus, the number of features has been increased from 41 to 118.

Two different multiagent system were organized with two data subsets selected from the 10% training data set: DSS1 and DSS2 with 500 and 169,000 connection records,

Table 1. Data distribution of different data subsets

Connection Category	10% Training	DSS1	DSS2	10% Test
Normal	97278	30416	53416	60593
DoS	391458	64299	110299	223298
Probe	4107	4107	4107	2377
R2L	1126	1126	1126	5993
U2R	52	52	52	39
Unknown	0	0	0	18729

respectively. Both data subsets contain the 22 attack types in addition to normal records. The distribution of the selected data mirrors the distribution of data in the 10% training data set, which has an irregular distribution. The distribution of training and testing data subsets used is given in Table 1, where the 'Unknown' connection category represents the connection records of the new attack types present in the testing data set. We chose 0.1 and 0.08 as values for parameters τ in both systems, generating architectures of 19 and 15 agents, respectively. Training results are shown in Table 2. Related works are usually interested in distinguish between attacks or normal records. However, we are also interested in classify an anomaly into its attack type. Thus, the detected rate are the attacks that were detected as attacks, the false positive rate are the normal connection records that were detected as attacks, and the identified rate are the connection records that were identified as their correct connection type.

Table 2. Training results for DSS1 and DSS2

Training Set	Detected(%)	False Positive(%)	Identified(%)
DSS1	92.53	0	60.4
DSS2	95.05	0.69	95

Table 3. Testing results with 311,029 records for DSS1 and DSS2

Training Set	Detected(%)	False Positive(%)	Identified(%)
DSS1	90.97	1.5	91.06
DSS2	91.07	3.53	90.52

After training, the two systems were tested with the 10% testing data set. Testing results are given in Table 3, where detection rates are similar but the false positive rate is lower for the first trained system. It can be owing to the high amount of training data for the second system, which yielded a more specialized system in those data, whereas the first system lernt more generally the structure patterns.

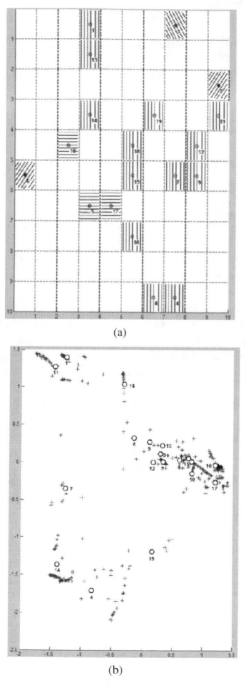

(a)

(b)

Fig. 4. Obtained architecture after training with DSS1 and $\tau = 0.1$: (a) Generated topology with 19 nodes, where nodes with the same line pattern represent the same group in the first level of the hierarchy. (b) Two principal components representation space with 19 nodes (circles) and 500 input data (crosses).

Table 4. Testing results for differents IDSs based on self-organization

	Detected(%)	False Positive(%)	Nodes
GSOM MAS	90.97	1.5	19
K-Map	99.63	0.34	144
SOM	97.31	0.042	400
SOM (DoS)	99.81	0.1	28800

Most of the related works have used self-organizing models to implement an IDS. In [8], a hierarchical Kohonen net (K-Map) is composed of three pre-specified levels, where each level is a single K-Map or SOM. Their best result was 99.63% detection rate after testing, but taking into account several limitations. Three attack types were just used during the testing, while we used 38 attack types, where 15 attack types were unknown. Also, they used a combination of 20 features that had to be established in advance, and 48 neurons were used in each level. From [3], we chose the only one SOM trained on all the 41 features in order to compare results. This neural network achieved a detection rate of 97.31%, but using 400 neurons. An emergent SOM (ESOM) for the Intrusion Detection process was proposed in [7]. They achieved detection rates between 98.3% and 99.81% and false positives between 2.9% and 0.1%. However, the intrusion detection was just limited to DoS attacks, they used a pre-selected subset of 9 features and between 160x180 and 180x200 neurons were used. In addition, their best result (99.81% detection rate and 0.1% false positive rate) was just trained and tested with one DoS attack type, the smurf attack. A comparison between our experimental results (GSOM Multiagent System) and the rest of related works results are summarized in Table 4.

Once the unsupervised learning process is finished, hierarchical relations among input data must be represented by the automatic generated architecture. For this purpose, the neural network is trained again, but this time using the weight vectors of nodes as input data. This way, the resulting nodes are self-organized in groups, so these groups of nodes indicate a two-level hierarchy of input data. The generated topology and the two principal components representation space of input data from DSS1 are shown in Fig. 4. Note that in Fig. 4(a), nodes have different line patterns, where nodes with the same pattern belong to the same group in the upper layer of the hierarchy. Nodes belonging to the same group are not necessary to be together in the topology, since they represent a group in the representation space that is not linearly separable.

5   Conclusions

A multiagent system based on a growing self-organizing map has been proposed in this paper. This system has capabilities to analyze and discover knowledge gathered from distribuited agents. The central administrator agent keeps a global knowledge that is accessible by the other agents for further information about the domain and controls the size of the architecture by means of clustering techniques based on the proposed growing SOM.

In order to faithfully represent input data, the proposed growing SOM provides a more flexible adaptation of its topology by adding or deleting nodes. Moreover, hierarchical relations among data are provided by clustering the generated nodes and showing a two-levels hierarchy of input data mapped in a 2-D representation space.

This novel multiagent system has been used to implement an Intrusion Detection System (IDS). It was trained and tested with the KDD Cup 1999 benchmark data set. After testing, we achieved a 90.97% detection rate and a false positive rate of 1.5%. These results are compared with the results from related works. Apparently these other results are better than ours, however they have several limitations in common. First, the number of features and attacks are reduced for training andor testing. Second, the model architecture has to be fixed in advance and is static. Third, hierarchical relations from data are not represented. Finally, the architecture complexity is higher due to the big amount of nodes used. All these observations indicate that a previous knowledge about problem domain is needed in these works, which is difficult in many problem domains, especially when high-dimensional data are present.

Acknowledgements

This work is partially supported by Spanish Ministry of Science and Innovation under contract TIN-07362, project name Self-Organizing Systems for Internet.

References

1. Cannady, J.: Artificial neural networks for misuse detection. In: anonymous (ed.) Proceedings of the 1998 National Information Systems Security Conference (NISSC 1998), Arlington, VA, October 5-8, pp. 443–456 (1998)
2. Czibula, G., Guran, A., Cojocar, G., Czibula, I.: Multiagent decision support systems based on supervised learning. In: IEEE International Conference on Automation, Quality and Testing, Robotics, 2008. AQTR 2008, vol. 3, pp. 353–358 (2008)
3. DeLooze, L.L.: Attack characterization and intrusion detection using an ensemble of self-organizing maps. In: 7th Annual IEEE Information Assurance Workshop, pp. 108–115 (2006)
4. Kohonen, T.: Self-organized formation of topologically correct feature maps. Biological cybernetics 43(1), 59–69 (1982)
5. Lee, W., Stolfo, S., Chan, P., Eskin, E., Fan, W., Miller, M., Hershkop, S., Zhang, J.: Real time data mining-based intrusion detection. In: DARPA Information Survivability Conference & Exposition II, vol. 1, pp. 89–100 (2001)
6. Maxion, R., Tan, K.: Anomaly detection in embedded systems. IEEE Transactions on Computers 51(2), 108–120 (2002)
7. Mitrokotsa, A., Douligeris, C.: Detecting denial of service attacks using emergent self-organizing maps. In: 5th IEEE International Symposium on Signal Processing and Information Technology, pp. 375–380 (2005)
8. Sarasamma, S., Zhu, Q., Huff, J.: Hierarchical kohonen net for anomaly detection in network security. IEEE Transactions on Systems Man and Cybernetics Part B-Cybernetics 35(2), 302–312 (2005)
9. Tan, K., Maxion, R.: Determining the operational limits of an anomaly-based intrusion detector. IEEE Journal on Selected Areas in Communications 21(1), 96–110 (2003)

10. Wickramasinghe, L., Alahakoon, L.: Dynamic self organizing maps for discovery and sharing of knowledge in multi agent systems. Web Intelli. and Agent Sys. 3(1), 31–47 (2005)
11. Ying, H., Feng, T.-J., Cao, J.-K., Ding, X.-Q., Zhou, Y.-H.: Research on some problems in the kohonen som algorithm. In: International Conference on Machine Learning and Cybernetics, vol. 3, pp. 1279–1282 (2002)
12. Zhang, W.-R., Zhang, L.: A multiagent data warehousing (madwh) and multiagent data mining (madm) approach to brain modeling and neurofuzzy control. Inf. Sci. Inf. Comput. Sci. 167(1-4), 109–127 (2004)

Towards Cooperative Predictive Data Mining in Competitive Environments

Viliam Lisý, Michal Jakob, Petr Benda, Štěpán Urban, and Michal Pěchouček

Agent Technology Center, Dept. of Cybernetics, FEE, Czech Technical University
Technická 2, 16627 Prague 6, Czech Republic
{lisy,jakob,benda,urban,pechoucek}@agents.felk.cvut.cz

Abstract. We study the problem of predictive data mining in a competitive multi-agent setting, in which each agent is assumed to have some partial knowledge required for correctly classifying a set of unlabelled examples. The agents are self-interested and therefore need to reason about the trade-offs between increasing their classification accuracy by collaborating with other agents and disclosing their private classification knowledge to other agents through such collaboration. We analyze the problem and propose a set of components which can enable cooperation in this otherwise competitive task. These components include measures for quantifying private knowledge disclosure, data-mining models suitable for multi-agent predictive data mining, and a set of strategies by which agents can improve their classification accuracy through collaboration. The overall framework and its individual components are validated on a synthetic experimental domain.

1 Introduction

We study the case of multiple self-interested parties (termed *agents* further on) working on a common but partitioned predictive data mining task[1] in a competitive environment. The knowledge used in solving the data mining task is considered a valuable asset of each party because accurately predicting data classifications is assumed to give the party a competitive advantage. One of the ways the classification accuracy can be improved is by exchanging knowledge with other parties. Given the competitive nature of the domain, such exchange cannot be performed on an arbitrary basis but needs to follow sound strategies that consider knowledge lost and gained during each transaction. The ability to implement such strategies is a necessary precondition to enabling cooperation between the agents.

Real world examples of the above setting are for example banks that use data mining models to decide whether to authorize a loan to a client. The models and their parameters in this case are well-kept secrets that help the banks to win more reliable clients and consequently increase their profits. At the same time, banks in this situation are willing to cooperate to some extent as indicated by the existence of shared registers of defaulters in many countries. However, more advanced cooperation strategies allowing

---

[1] Different agents have different datasets but all the datasets are sampled from a single underlying model.

L. Cao et al. (Eds.): ADMI 2009, LNCS 5680, pp. 95–108, 2009.

the control over the amount of disclosed private knowledge could enable and motivate more efficient cooperation beneficial for all parties.

In this paper, we outline a framework and describe a set of specific methods and techniques that make the implementation of such advanced strategies possible. First, we develop a set of private knowledge disclosure metrics that measure the classification knowledge lost in data and model exchanges between the agents. We identify the set of operations that need to be supported by the underlying classification models to make them applicable in a cooperative setting, in particular the possibility to effectively evaluate private knowledge metrics and to incrementally accommodate prediction knowledge obtained from other agents and, vice versa, to effectively extract knowledge to be shared. Not all classification models fulfil these requirements – we identify the Naïve Bayes classifier as a particularly suitable class of models for semi-cooperative prediction, and we show how individual operations can effectively be implemented for this class. We then describe several cooperative strategies that the agents can use to improve their classification capability. For each strategy, we measure the resulting improvement of agent's classification capability and set it against the loss of privacy entailed.

The work presented should be viewed as an initial step towards enabling cooperation in predictive data mining in a community of self-interested and competitive agents.

2 Related Work

Over the last decade, privacy preserving data mining (PPDM) [1] has attracted considerable attention. One of the main objectives of the field is designing data transformations which allow publishing data without losing privacy. In most cases, the emphasis is on protecting the privacy of *individuals* described by the data records – not on the protection of knowledge contained in the data set as a whole. Consequently, losing exact information about a small number of records is considered unacceptable while complete models learned from the data can be freely shared unless they allow the derivation of individual records [2].

This is well acceptable for the preservation of private information about individuals, such as medical records, but it is not sufficient in reasoning about the preservation of private data, knowledge, and know-how of businesses. Almost any data mining result, a cluster model, a frequent sequence, or an association rule derived from private company data can have commercial potential and should not be blithely shared without further assessment.

The subfield of PPDM that deals with this kind of private information, i.e. the knowledge contained in collections of data rather than individual records, is termed *corporate privacy* in [3] or *knowledge hiding* e.g. in [4]. There is, however, no general framework to reason about private knowledge disclosure – each technique has its own methods and measures of quality of private knowledge hiding. The research closest to the focus of this paper is classification rules [5] and association rules [6] hiding. The objective of these methods is to transform data by resampling, reduction and/or perturbation so that a specified set of rules cannot be mined from the data; the outcome is measured in terms of the number of rules successfully hidden, the number of original data items modified and the number of new rules introduced by the transformation.

In contrast to the above, in this paper, we do not aim to conceal any specific classification rules; instead, we aim to measure and minimize the overall knowledge disclosed about private classification models possessed by individual agents (and possibly obtained by generalizing their private data). As far as measures for quantifying privacy loss in PPDM are concerned, a summary is presented in [7]. Existing measures, often utilizing the notion of information entropy, are designed to measure the disclosure of a specific kind of knowledge. Moreover, except for a few exceptions (e.g. [8] for association rules), existing measures deal with individual privacy and are therefore not applicable to our problem.

On the other hand, privacy loss measured in terms of changes to information entropy has been used in multi-agent system research outside the data mining field (e.g. in multi-agent meeting scheduling [9], or multi-agent planning [10]).

In order to design the interaction strategies, the first step is to assign utility to privacy loss, classification precision gain and other important parameters of the task. A similar problem is pursued in utility-based PPDM [11], but this general problem and approach need to be revisited with respect to the specific kind of private knowledge we consider. Furthermore, once an agent's preferences over its possible strategies and the preferences of the other agents in the system are defined, the agent can use the game theoretical framework for rational decision-making about privacy loss [12].

3   Problem Description

The problem addressed is *semi-cooperative classification* in competitive multi-agent domains. Each agent aims to accurately classify a set of unlabelled examples drawn from the same data distribution. In order to improve its classification accuracy, the agent can either exploit its own data and models or request additional knowledge from other agents. More specifically, the problem is defined as follows. There are multiple agents in the system, each containing its:

- set of labelled training data D_i
- set of unlabeled data for classification C_i (termed *task data*)
- data mining algorithm M_i that can construct a classification model $M_i(S)$ from any labelled set of data S

The task of each agent is to classify its unlabelled task data C_i. In doing so, the agent employs a particular *strategy* which involves exploiting its training data D_i and communicating with other agents about their models, data and classifications. Each strategy results in a specific classification accuracy (measured by a designated accuracy metric – see Section 4.1) and in disclosure of a specific amount of private knowledge sourced externally (measured by a designated private knowledge loss metric – Section 4.2).

4   Relevant Measures

In this section, we introduce measures for evaluating agent's knowledge sharing actions with respect to (1) the loss of private knowledge and (2) the increase in classification accuracy. We briefly introduce the measures here and discuss their properties and suitability in more detail in Section 7.

4.1 Private Knowledge Loss

According to our literature survey, no metrics for measuring the loss of information about a private classification model has been so far reported. In the following, we describe and analyze how symmetrised Kullback-Leibler divergence and mutual information can be used for this purpose. The application of these measures to quantify private knowledge loss is novel.

Both proposed measures are based on the following idea. If an agent sends some information originating from its private classification model (e.g. classified examples or partial models) to another agent, this information can be used to (approximately) reconstruct the private classification model of the sending agent. For example, if an agent discloses a set of feature vectors classified by the private model, these samples can be used to train a new model that approximates the original one. The proposed measures therefore evaluate the distance between the original private model and the (hypothetical) reconstructed model and use it as a quantification of the private classification knowledge loss.

The proposed measures assume that any classifier can be interpreted in two ways. Either as a representation of a joint probability distribution over the feature space and the classification labels, or as a realization of a function assigning a unique label to each feature vector. Although the first representation is more informative, it is often hard to extract from certain classes of classification models. For example, the k-nearest neighbour classifier approximates the distribution by the ratio of examples of individual classes in the neighbourhood of the feature vector; decision trees can approximate the distribution for a feature vector by the ratio of class frequencies of training examples corresponding to the leaf the feature vector belongs to. For the cases where the distribution can be extracted, we propose a measure based on the probabilistic similarity measure between the original distribution represented by the private classifier and the distribution realized by the reconstructed classifier.

On the other hand, extracting the distribution from a set of rules derived by ILP or similar method is not straightforward. Moreover, the internal structure of some classifiers is not always accessible to the agent (e.g. in the case of external libraries) and the classifier can be accessed only as a black-box producing a classification (class label) for any input feature vector. If classifications are the only available information about the classifier, we propose a metric based on information theory that measures the information present in the classifications produced by the reconstructed model about the classifications corresponding to the original private model.

Symmetrized Kullback-Leibler Divergence. Kullback-Leibler (KL) divergence (also termed *information divergence* or *information gain*) is a standard measure used in probability theory to compute the similarity of two probability distributions. For two discrete probability distributions P and Q over the same random variable (the same feature space in our case), the KL divergence of Q from P is defined as

$$D_{kl}(P\|Q) = \sum_x P(x) log_2 \frac{P(x)}{Q(x)} \tag{1}$$

where x in the sum iterates over the range of the random variable. KL divergence in this form is not symmetric, i.e., $D_{kl}(P\|Q) \neq D_{kl}(Q\|P)$ in most cases. That is why the symmetrised form of Kullback-Leibler divergence is often used

$$D_{skl} = D_{kl}(P\|Q) + D_{kl}(Q\|P) \tag{2}$$

The iteration over the range of the random variable corresponds to iterating over the whole feature space in the case of a distribution realizing a classifier. This is generally computationally expensive for large distributions. For example in our case of 10 features with 10 possible values each, the size of the feature space is 10^{10}. However, for some specific classifiers, the value of KL-divergence can be approximated or even exactly computed in a significantly more effective way. (See Section 5.2).

Mutual Information. Mutual information is a standard information-theoretic measure quantifying the mutual dependence of two random variables. It represents the amount of uncertainty about one random variable that is removed by knowing the value of the other random variable. This measure can be used to measure how much information a model reveals about a data set or about the classifications produced by a model regardless of the structure of the classifier.

For random variables X and Y, mutual information is defined as

$$I(X:Y) = \sum_{x \in X} \sum_{y \in Y} P(x,y) log_2 \frac{P(x,y)}{P(x)P(y)} \tag{3}$$

Mutual information can be used to compute both the amount of information a model reveals about a data set and the amount of information one model contains about another model. We focus on the second case here. Classifications produced by the private model and by its approximation correspond to the two random variables in the formula. The only step needed is to estimate the probability distribution $P(X,Y)$, i.e., the normalized matrix specifying how frequently one classifier classifies a feature vector to class x while the other classifies the same vector to class y, which may or may not be the same. In the case of a smaller feature space, this probability distribution can be computed exactly by iterating over the whole feature space, similarly to the case of the distribution-based metric. Such explicit computation is not possible for large feature spaces. Instead, we therefore sample the two classification models on a chosen subset of the feature space and estimate the probability distribution $P(X,Y)$ from the confusion matrix resulting from the two sets of produced classifications.

A big advantage of this measure is its complete independence on the classifier implementation. A disadvantage of the mutual information metric is the amount of samples needed for good approximation of the distribution. We have performed a simple synthetic experiment that demonstrates this problem and shows some basic properties of the metric. We have simulated classification into 10 classes as in the experiments presented later. We were measuring the information that an imprecise classifier contains about an ideal classifier that always assigns the correct class. The probabilities in formula 3 can be estimated from the confusion matrix of the classifier, so we were generating only that matrix in this experiment. Correct classification was provided with probability p,

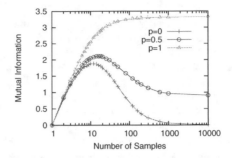

Fig. 1. Dependence of the mutual information metric on the number of samples used to construct the confusion matrix (from which the metric is calculated) for different probabilities of correct classification

otherwise the classification was random to any of the remaining classes with uniform distribution. Figure 1 shows the dependence of the mutual information metric on the number of generated classifications in the confusion matrix in a single run of the experiment. The probability of correct classification were set to $p = 0$, $p = 0.5$ and $p = 1$, respectively, and the results shown are the mean of thousand runs. The variance was very high for less than 100 samples but decreased for more samples; the results obtained from more than 1000 samples were already almost identical across individual runs.

As we expected, when there is enough data samples, the mutual information between the correct classification and a classifier that always misclassifies into one of the nine wrong classes is almost zero. If the classifier classifies a half of the examples correctly, the mutual information is higher, and it is the highest when the classifier classifies all examples correctly. In the latter case of correct classification, it converges to the value of 3.32 bits, which corresponds to distribution $P(X,Y)$ with 10 classes and the probability uniformly distributed along the main diagonal.

The inaccurate probability estimations caused by a smaller number of samples lead, besides high variance, to a misleading increase in the measured information disclosure e.g. incorrectly indicating that ten random classifications reveal more information about the private model than a thousand classifications that are correct in half of the cases. In order to obtain reliable assessment of information disclosure using the mutual information metric, at least a thousand samples are needed.

4.2   Classification Accuracy

There are many measures for assessing classification accuracy. Some of them are based on ratios of true positive and true negative classifications or consider different costs of misclassification to different classes – see e.g. [13] for an overview. The primary focus of this paper is on private knowledge preservation. We therefore use a ratio of correctly classified examples to all examples on a test data set as a basic classification accuracy measure.

5   Naïve Bayes for Multi-agent Data Mining

A data mining model suitable for semi-cooperative multi-agent classification task should satisfy several criteria:

- *creating and joining sub-models.* If agents exchange sub-models, their creation and joining should be possible and computationally efficient.
- *confidence.* The model should be able to output the degree of confidence in the classification to allow the agents to decide when they need some extra information from other agents.
- *effective measures computation.* The model should allow fast computation (or approximation) of relevant measures, in particular the private knowledge loss.
- *compact model size.* The models should have a compact representation to allow sharing in case of limited communication bandwidth.

One of the models that satisfies all the above properties is Naïve Bayes classifier. Joining two models is straightforward if they are represented as frequencies instead of probabilities (see Section 5.2). Creating a model requires only one iteration through the data set. The classifier output in the form of a posterior probability is suitable confidence measure and as we show below, the suggested privacy measures can be computed exactly in a reasonable time for this model.

5.1   Naïve Bayes Classifier

Let X is an n-dimensional feature space, X_i is a random variable representing the i-th component of the feature vector, and S is a set of all classes. Naive Bayes then classifies a given sample $(x_1,\ldots,x_n)$ into class

$$d = \arg\max_{s\in S}\left[P(S=s)\prod_{i=1}^{n}P(X_i=x_i|S=s)\right] \tag{4}$$

The product in the formula approximates the joint probability

$$P(X_1=x_1,X_2=x_2,\ldots,X_n=x_n|S=s) \tag{5}$$

under the assumption of the conditional independence of individual features.

5.2   Joining Naïve Bayes Models

The model sharing strategies require the ability to join classification models. In our framework, this is implemented in the following way. Naïve Bayes classifier is composed from a set of probability distributions

$$M = \bigcup_{s\in S}\{P(s),P_1(x_1|s),\ldots,P_n(x_n|s)\} \tag{6}$$

where $x_1\in X_1,\ldots,x_n\in X_n$. We have implemented the model as

$$\bigcup_{s\in S}\{f(s),f_1(x_1|s),\ldots,f_n(x_n|s)\} \tag{7}$$

where $f(s)$ is the frequency of class s in the training set and $f_i(x_i|s)$ is the frequency of i-th attribute in the subset of the training data corresponding to class s. The number of data samples in the training set used to train the classifier can be directly determined from the implemented model as $a = \Sigma f(s)$. Probability $P(s)$ is then given by $\frac{f(s)}{a}$ and probabilities $P_i(x_i|s) = \frac{f_i(x_i|s)}{f(s)}$

Two models M^1 and M^2 are then joined as:

$$\bigcup_{s \in S} \{f^1(s) + f^2(s), f_1^1(x_1|s) + f_1^2(x_1|s), \ldots, f_n^1(x_n|s) + f_1^2(x_n|s)\} \tag{8}$$

KL-Divergence for Naïve Bayes. The Naïve Bayes classifier and its feature independence assumption allows to significantly simplify the computation of certain measures. A simplified calculation of the otherwise computationally expensive KL-Divergence for two probabilistic distributions P_1 and P_2 representing two Naïve Bayes classifiers is shown below[2]

$$D_{kl}(P_1\|P_2) \quad = \quad \sum_{s \in S} P_1(s) \left(log_2 \frac{P_1(s)}{P_2(s)} + \sum_{i=1}^n \sum_{x_i \in X_i} P_1(x_i|s) log_2 \frac{P_1(x_i|s)}{P_2(x_i|s)} \right) \tag{9}$$

where n is the number of features, X_i are the domains of the features and S is it set of classes.

6 Cooperation Strategies

As described in Section 3, the agents can employ a cooperation strategy in order to improve their classification accuracy. The strategy can either involve requesting information about other agents' private models or asking the other agents to classify a set of unlabelled data using their private models. In the following description of the strategies and their schemas in Figure 2, we use the term *requestor* for the agent aiming to improve its classification by utilizing the private knowledge of the other agents (which are termed *providers*).

6.1 Model and Sub-model Sharing

In this strategy, the requestor asks the providers for a partial or degraded version of their private models. In our experiments, the degraded version is a model trained only on a selected amount of samples randomly taken from the training set of the provider. Other means of degradation, such as adding noise to the model, or defining the model only on a subset of the feature-space, can be employed. The providers willing to share some information create partial models degraded to a level satisfying their private knowledge disclosure constraints and send them to the requestor. The requestor then merges the information provided in the received partial models into its own private model and uses the resulting improved model to classify its task data. In the case of the Naïve Bayes model, the merge algorithm has been described in Section 5.2.

---

[2] Detailed derivation is available on request from the authors.

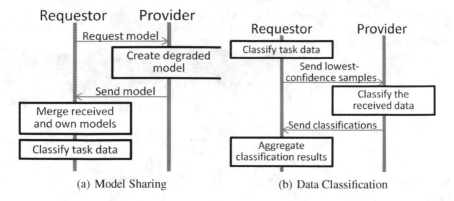

Fig. 2. Schemas of the cooperation strategies

6.2 Data Classification

In the case of the data classification strategy, the requestor first classifies its task data using its private classifier and sorts the results according to the confidence of the classification (the a posterior probability in the case of the Naive Bayes model). It then decides about the fraction of data with the lowest classification confidence and sends them to the providers for classification. The providers use their private classification models to label the data and send the resulting classifications back to the requestor. The requestor aggregates the received classifications with the results of its own classifier and determines the final classification of the data. The aggregation mechanism used in our implementation is majority voting with a preference for own classification in case of ties. A certain disadvantage of the data classification strategy for the requestor is that by specifying the data to be classified the requestor discloses information about its own data.

In order to measure the private knowledge loss using the KL measure, an additional step is needed when using the data classified strategy. The labelled data provided by the providers in response to the requestor's queries are first used to train a Bayes classifier. The classifier approximates the private model of the provider and it is then used to calculate the amount of private knowledge disclosed.

7 Experiments

In order to validate our approach and to obtain quantitative data on the behaviour of the proposed metrics and strategies, we have designed a set of experiments on a synthetic domain.

7.1 Experiment Domain and Setting

The feature space is composed of two features; each feature can assume a value from $\{0, 1, \ldots, 9\}$. The feature vectors belong to one of ten possible classes with equal probability. Data samples for each class are drawn from a Gaussian distribution (restricted to

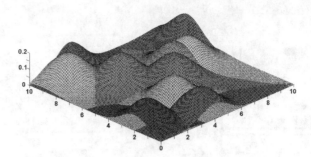

Fig. 3. The domain used in the experiments. Each Gaussian generates data samples from a different class.

the feature space) defined by its mean and a fixed unit variance. The distribution means are selected randomly with a uniform probability distribution from the whole feature space. As a result, identical feature vectors with different classifications can appear in the domain. The situation is illustrated in Figure 3.

The community of agents in the experiments comprises ten identical agents $A_0, A_1, \ldots A_9$. Each agent employs one of the two cooperation strategies (see Section 6) to classify its own task data with the help of the classification knowledge obtained from all other agents. Each agent has its private classification model trained on hundred data samples (D_i) and it has another thousand data samples to classify as its task data (T_i). All presented results are averages over one hundred runs of the experiment with identical settings.

KL divergence as well as the naïve Bayes classification works well only with positive distributions, i.e. each combination of features and classification is considered possible (there are no zeros in the distribution). In order to achieve this property in our experiments, we initialize the frequencies representing the Naïve Bayes classifier as explained in Section 5.2 with ones. However, these extra ones are kept only once in the process of model merging.

7.2 Results for the Model Sharing Strategy

The main parameter of the model sharing strategy is the portion of the training data the provider uses to train a degraded classifier that it sends to the requestor. The dependency of individual measures on this parameter are depicted in the upper row of graphs in Figure 4.

Classification Accuracy. Classification accuracy measure plotted in Figure 4(a) shows how model parts from other agents merged together with the original model of the requestor agent A_0 improve requestor's classification accuracy. The increase in prediction accuracy was expected – the bigger the part of provider agents' models merged into the model of the requestor, the more information about the domain the resulting model captures. If the requestor A_0 receives models learned only from 5% of providers' data, the accuracy of its classification is approximately 61%. The accuracy increases quickly at

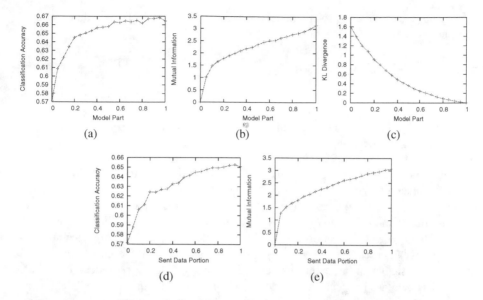

Fig. 4. Classification precision of requestor A_0 using the model sharing strategy and the private knowledge loss of agent A_1 responding to requestor's queries according to various measures for the model sharing (a,b,c) and the data classification (d,e) strategies

the beginning but the improvement tails off closer to sharing full models; this is because the private classification model is already well approximated at this point.

Private Knowledge Loss. In this strategy, the provider agents disclose their private knowledge in the form of partial classification models – see the results for the individual measures below. Note that the requestor agent does not disclose any information at all.

Mutual Information. The mutual information is measured between the original private model of the provider and the (degraded) model sent to the requestor. The data used to sample the distribution $P(X,Y)$ from formula 3 are the complete training set the provider used to obtain its private classifier. The dependency of the mutual information between the models and the portion of the data used to train the shared model is depicted in Figure 4(b). Even though it is measured on a different agent, the trend is similar to the classification improvement of the requestor. It means that the mutual information metric accurately describes the amount of useful information contained in the sent model; more data used for creating the degraded model implies higher private knowledge disclosure.

KL-Divergence. Figure 4(c) depicts the trend of KL divergence, which measures the difference between the disclosed model and the private one. The trend is in agreement with expectations – initially, increasing the amount of data on which the degraded model is created contributes strongly to narrowing the difference between the provider's private model and the sent model; the contribution decreases when the degraded model approaches the provider's private model (i.e. when the portion of the training data used approaches one).

7.3   Results for the Data Classification Strategy

The main parameter of the data classification strategy is the percentage of the task data with the smallest classification confidence the requestor sends to the providers for classification. The experimental results concerning this strategy are summarized in the bottom row of graphs in Figure 4.

Classification Accuracy. Figure 4(d) confirms the basic hypothesis that the classification accuracy can be improved using the data classification strategy – the accuracy grows clearly with the increased amount of shared testing data of the requestor. The irregularities in the trend are caused by fluctuations of the classification precision between individual runs – a smoother graph would be obtained for a higher number of runs or a large data set. The improvement through collaboration is largest when the data on which the requestor has a very low confidence are sent to the providers for classification. The increase in accuracy diminishes when the agent requests classifications for the majority of its testing data; this is because then the requestor already has a good chance of predicting the class correctly.

Private Knowledge Loss. In contrast to the model sharing strategy, both the requestor and the providers lose private knowledge in this case. The requestor A_0 loses private knowledge about its task data while the provider agents lose information about their private models.

The amount of information lost by the requestor can be quantified by the number of examples it sends to the other agents for classification. The loss for the providers can be expressed by the proposed measures.

Mutual Information. The plot in Figure 4(e) depicts the loss of knowledge about providers' private models caused by responding to requester's queries. The loss of the private knowledge of providers increases with the number of queries answered, fast at the beginning and slower towards the end when the requestor already has enough information to reconstruct the providers' models accurately.

KL-Divergence. We do not present a plot of KL divergence metric for this strategy because the metric proved unsuitable for the data classification strategy. KL divergence computes the distance between two probability distributions realized by the corresponding classifiers. In order to use KL divergence, we first need to reconstruct the classifier from the classified samples sent. The problem arises from the fact that different distributions can correspond to identical classifications. If a feature vector can belong to multiple classes in the ground truth, the original classifier can learn this fact but it always outputs only the most probable class. A classifier trained only from sampling the original classifier will have the probability of classifying to the dominating class equal to one, and the probabilities of classifying to the other classes equal to zero. As a result, the KL divergence can grow even though the similarity of produced classifications increases.

7.4 Strategy Comparison

Both the evaluated strategies are usable for improving classification accuracy of the requestor agents. Starting from the same base level (57%), the model sharing strategy achieved up to 67% accuracy while the data sharing reaches only 65% even if all the thousand task data samples are shared. On the other hand, the model sharing strategy has to reveal the whole model of the provider agent (instead of classified samples) in order to achieve the peak accuracy. We compare loss of the private knowledge about the task data and the private knowledge contained in classifiers.

The information about the task data is communicated only in the case of data sharing strategy. The requestor agent sends a portion of its testing data given by the strategy parameter to all provider agents. The strategy parameter represents how much of the private testing data will be disclosed.

The knowledge about private classification model is disclosed in both strategies. The measure suitable for measuring the amount of private knowledge revealed in both of them is the mutual information metric. The graphs for both strategies are almost identical. According to that, sending model trained on k samples form an agent training set reveals as much private knowledge about the complete model as answering $10 * k$ queries for classification of other agents task data. The size of the task data sets is ten times bigger than the size of the training sets.

The inapplicability of KL divergence as a sound private knowledge loss measure for the data classification strategy indicates that the knowledge exchanged in the strategy does not allow to fully reconstruct a provider's model, in particular its classification confidence. This is not the case for the model sharing strategy where the confidence information is disclosed as part of the exchanged partial models.

8 Conclusion

We have analyzed the trade-offs between improving classification accuracy and losing private classification knowledge in a competitive multi-agent setting. Reasoning about such trade-offs is of high relevance whenever multiple competitive parties want to cooperate in order to improve their performance. We believe such situations are common in real-world environments, including various business sectors (e.g. banking, insurance etc) or areas of international cooperation.

Understanding that cooperation cannot arise unless individual parties can reason about losses and benefits entailed by such cooperation, we have designed a set of measures and a benchmark task on which the measures can be validated. Afterwards, we have presented two elementary strategies consisting of a set of operations which are likely to be the building blocks of any multi-agent strategy aimed at improving classification accuracy through agent cooperation. The first strategy is based on sharing partial classification models; the other strategy employs consulting other agent's opinions about the classification of specific examples. We have evaluated the strategies on a synthetic domain and compared their respective advantages and disadvantages. The results obtained are in agreement with our theoretical expectations and indicate that the framework developed could be used for building semi-cooperative data mining systems.

The work presented is, in any case, only a first step towards enabling more complex strategies e.g. involving reciprocity or payment for the private knowledge disclosed. In the longer term, we aim to design fully autonomous agents that can automatically reason about the benefits and costs of individual knowledge sharing operations in the context of semi-cooperative predictive data mining.

Acknowledgments

We gratefully acknowledge the support of the presented research by Office for Naval Research project N00014-09-1-0537 and by the Czech Ministry of Education, Youth and Sports under Research Programme No. MSM6840770038: Decision Making and Control for Manufacturing III.

References

1. Agrawal, R., Srikant, R.: Privacy-preserving data mining. In: Proc. of the ACM SIGMOD Conference on Management of Data, pp. 439–450. ACM Press, New York (2000)
2. Kantarcioğlu, M., Jin, J., Clifton, C.: When do data mining results violate privacy? In: KDD 2004: Proceedings of the tenth ACM SIGKDD international conference on Knowledge discovery and data mining, pp. 599–604. ACM, New York (2004)
3. Clifton, C., Kantarcioglu, M., Vaidya, J.: Defining Privacy for Data Mining. In: National Science Foundation Workshop on Next Generation Data Mining, pp. 126–133 (2002)
4. Bonchi, F., Saygin, Y., Verykios, V., Atzori, M., Gkoulalas-Divanis, A., Kaya, S., Savas, E.: Privacy in Spatiotemporal Data Mining. In: Mobility, Data Mining and Privacy: Geographic Knowledge Discovery (2008)
5. Natwichai, J., Li, X., Orlowska, M.: Hiding Classification Rules for Data Sharing with Privacy Preservation. In: Tjoa, A.M., Trujillo, J. (eds.) DaWaK 2005. LNCS, vol. 3589, pp. 468–477. Springer, Heidelberg (2005)
6. Verykios, V.S., Gkoulalas-Divanis, A.: A Survey of Association Rule Hiding Methods for Privacy. In: Privacy-Preserving Data Mining. Springer, US (2008)
7. Bertino, E., Lin, D., Jiang, W.: A Survey of Quantification of Privacy Preserving Data Mining Algorithms. In: Privacy-Preserving Data Mining. Springer, US (2008)
8. Bertino, E., Fovino, I., Provenza, L.: A Framework for Evaluating Privacy Preserving Data Mining Algorithms. Data Mining and Knowledge Discovery 11(2), 121–154 (2005)
9. Franzin, M., Rossi, F., Freuder, E., Wallace, R.: Multi-Agent Constraint Systems with Preferences: Efficiency, Solution Quality, and Privacy Loss. Computational Intelligence 20(2), 264–286 (2004)
10. van der Krogt, R.: Privacy loss in classical multiagent planning. In: IEEE/WIC/ACM International Conference on Intelligent Agent Technology, pp. 168–174 (2007)
11. Weiss, G., Saar-Tsechansky, M., Zadrozny, B.: Report on UBDM-05: Workshop on Utility-Based Data Mining. ACM SIGKDD Explorations Newsletter 7(2), 145–147 (2005)
12. Kargupta, H., Das, K., Liu, K.: A game theoretic approach toward multi-party privacy-preserving distributed data mining. In: Communication (2007)
13. Mitchell, T.M.: Machine Learning. McGraw-Hill, New York (1997)

Part III

Data Mining Driven Agents

Improving Agent Bidding in Power Stock Markets through a Data Mining Enhanced Agent Platform

Anthony C. Chrysopoulos, Andreas L. Symeonidis, and Pericles A. Mitkas

Department of Electrical & Computer Engineering
Aristotle University of Thessaloniki,
GR-54 124, Thessaloniki, Greece
anchrys@ee.auth.gr, asymeon@eng.auth.gr, mitkas@eng.auth.gr

Abstract. Like in any other auctioning environment, entities participating in Power Stock Markets have to compete against other in order to maximize own revenue. Towards the satisfaction of their goal, these entities (agents - human or software ones) may adopt different types of strategies - from na?ve to extremely complex ones - in order to identify the most profitable goods compilation, the appropriate price to buy or sell etc, always under time pressure and auction environment constraints. Decisions become even more difficult to make in case one takes the vast volumes of historical data available into account: goods' prices, market fluctuations, bidding habits and buying opportunities. Within the context of this paper we present Cassandra, a multi-agent platform that exploits data mining, in order to extract efficient models for predicting Power Settlement prices and Power Load values in typical Day-ahead Power markets. The functionality of Cassandra is discussed, while focus is given on the bidding mechanism of Cassandra's agents, and the way data mining analysis is performed in order to generate the optimal forecasting models. Cassandra has been tested in a real-world scenario, with data derived from the Greek Energy Stock market.

Keywords: Software Agents, Data Mining, Energy Stock Markets, Regression Methods.

1 Introduction

Agent technology has already proven its potential in various aspects of real-world trading and electronic markets. In complex environments, such as (Power) Stock Markets, where bipartite relationships between involved actors demand negotiation in order to come to an agreement, the utilization of autonomous and intelligent agents has become imminent [1, 2]. Numerous approaches have been employed in order to develop the optimal agent strategy with respect to the challenges the agents face. Among all issues rising when designing and developing systems for such highly dynamic markets, the extreme rate that data is generated at is considered of high importance. Additionally, like in all auction environments, decisions have to be made under extreme time pressure, making it difficult to apply simple algorithms and/or analytic strategies. These factors indicate that Data Mining (DM) could be a suitable technology for achieving an "intelligent" and efficient software solution.

L. Cao et al. (Eds.): ADMI 2009, LNCS 5680, pp. 111–125, 2009.
© Springer-Verlag Berlin Heidelberg 2009

Within the context of our work we provide a flexible, robust and powerful tool for dealing with all the issues related to the hyperactive and continuously changing Power Stock Market. We have built a multi-agent system (MAS) capable of efficiently handling the deluge of available data and of practicing various DM methodologies, in order to reach what seems to be an efficient prediction of the prices of goods of interest. Our platform was benchmarked on data provided by the Greek Power Stock Market, which is a dynamic, partially observable environment. This environment allows for different strategic approaches to be followed, while it is highly challenging, due to the fact that each decision made affects instantly the future moves or decisions of the platform and the Stock Market itself.

Looking at the bigger picture, one may argue that an agent developed can employ DM, in order to extract useful nuggets of knowledge that could give him/her a predictive advantage over other competitors. In this context, we have applied DM in order to: a) analyze the historical data from the past auctions in order to predict the future values in goods of interest and, b) induce market behavior models and incorporate them into or agents, in order to provide them with a predictive edge over the competition.

The rest of the paper is organized as follows: Section 2 provides an overview of the Power Stock Market mechanisms (Auctions and Energy Market) available, as well as a state-of-the-art analysis. Section 3 presents the platform developed for monitoring and participating in the Power Stock market, and briefly discusses its architecture. Section 4 describes in detail the DM methodology applied, in order to decide on the optimal forecasting bid model, while Section 5 provides a pilot case scenario, aiming to illustrate the functionality of our platform. Finally, Section 6 summarizes work conducted and concludes the paper.

2   Power Markets

Up until recently, in most EU countries (Greece included), power supply was a physical monopoly, thus the establishment of a state or state-controlled administration department that would be responsible for producing, transferring and distributing, was justified. However, the advancement to more loose economic competition models has signified the cease of this physical monopoly, as far as the production and the distribution of energy are concerned. In turn, the liberation of the Power Market has given room for the development of Open Markets, where participants are able to choose between different energy products in different periods of time and may negotiate on their "folder of products". These folders can be negotiated under three different schemes:

- **The Long-term Market**, where participants come to direct agreements in form of long-term contracts.
- **The Day-ahead Market**, where buyers place their bids in 24 hourly auctions, in order to establish a contract for the next day.
- **The Real-time Market**, where buyers place their bids in order to establish a contract for the next hour.

Due to the physical model of the energy production-distribution-consumption network, long-term and day-ahead markets are the most dominant ones. Nevertheless, the inability to store Power for long periods of time, dictates the development of a mechanism that can efficiently balance the supply and demand of Power and can be easily and constantly controlled. The administrator of each Power system is responsible for ensuring this balance between production and consumption, through the utilization of a Real-Time Market. The Real-Time Market model bears the highest risk for the participants, since the malfunction of an Electric node or the shutting down of a Power line can bring exaggerated rising or falling of the prices. Nevertheless, its existence is decisive for the coverage of Demand, in case the other two Markets do not provide enough Power. In terms with these facts, we should add that Real-time Market involves the higher risk but also implies high profit maximization for players willing to participate in this Market.

2.1 Power Market Auctions

An *auction* is defined as a strict set of rules for the specification of the exchange conditions of goods [3]. In each auction, a number of transactions are performed between the participants, where each transaction comprises two elements: a) a protocol that defines the rules of the transaction mechanism as well as the actions allowed to an (human or software) agent participating in an auction and, b) a strategy, i.e. the methodology followed by an agent in order to fulfill its goal. The protocol of an auction is determined during its design and is announced to all the participants from the beginning. Agents' strategy is designed by each participant and is unknown to the rest.

In *Power Market Auctions*, two are the most important entities:

1. The Market participants (or Players)
2. The Independent System Administrator (ISA)

A *Player* is defined as any economical entity that accesses the Power Market [4]. In general, this entity may possess a group of Production Units or/and a group of Consumers. Each Player participating in the Power Market as a *Producer* should submit his/her power supply offers in pre-specified time intervals. Each offer contains the amount of supplying Power, as well as the minimum price he/she is willing to accept. One the other hand, each Player that participates in the Power Market as a *Consumer* should submit his/her power supply demands - within the same time intervals - along with the maximum price he/she is willing to pay for it.

The *ISA* is the administrator of the Power Transfer System, and also the Administrator of the Power Stock Market. Thus, ISA is responsible for the Settlement of the Power Market, taking into consideration the transferring limitations of the system. ISA collects bids for each hourly auction and has to calculate two curves: the aggregate ascending Supply Curve and the aggregate descending Demand Curve. In the simplified case where no transfer limitations are presented, the Settlement of the Market is the intersections of the two curves (Figure 1). This point determines the *Settlement Price of the Market* (SPM) - this is the price that the Producers are paid by the Consumers, the load production of each Production Unit and the consumption of each Consumer.

In case transfer limitations exist, ISA must follow a more complicated process. He/she then has to solve an optimization problem targeting to the maximization of the social prosperity (target function). Then the Power price, the Load production of each Production Unit and the consumption of each Consumer are calculated.

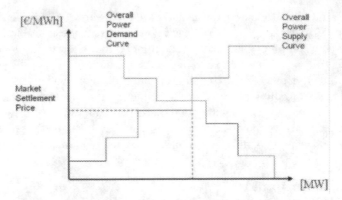

Fig. 1. The Aggregate Power Supply And Demand Curves

2.2 State-of-the-Art

Various approaches have been employed for analyzing the behavior of Power Markets, some of which have adopted Agent Technology (AT) and DM primitives. In fact, results reported seem quite promising.

In the early stages, the few implementations specifically designed for Energy Markets were merely simulations for the regulations or didn't make use of the advantages provided by the utilization of DM methodology. One of the first platforms created was MASCEM (Multi-Agent Simulator for Competitive Electricity Markets) [5] - [8], which were used to validate regulations and behaviors within the Electricity Markets, using only naive reinforcement learning strategies. Another Agent platform was soon developed in order to test the reliability of the proposals FERC of USA (Federal Energy Regulatory Commission) applied in the Standard Market Design (SDM) [9]-[10]. In this implementation, the Producers (Enterprises or Persons), the Consumers and the ISA were modeled by Agents.

In the early 2000s , during the bloom of MAS utilization of DM technology, the Electric Power research Institute (ERPI) developed SEPIA (Simulator for Electrical Power Industry Agents), a multi-agent platform capable of running a plethora of computing experiments for many different market scenarios [11, 12]. SEPIA employs two different options for agent learning: a variation of the Q-Learning Algorithm [13], which corresponds each noticeable state to a suitable action, and an LCS (Learning Classifier System) morph [14], which utilizes a rule-based model and agents learn through amplified learning and genetic algorithms.

The Argonne National Laboratory, on the other hand, developed EMCAS (Electricity Market Complex Adaptive System) [15], an efficient implementation for handling the Electric Energy Market. Through EMCAS, one may study the complex interactions between the physical entities of the market, in order to analyze the participants and their strategies. Players' learning is based on genetic algorithms, while EMCAS supports stock market transactions, as well as bipartite contracts.

As far as the real-market analysis is concerned, Bagnall [16] has presented a simplified simulation model of Great Britain's Electricity Market, were the Producers are agents that participate in a series of auctions-games. In any phase of the process, a Producer is faced with two options: a) try an already tested and applicable strategy that will reassure he/she will not lose money or, b) find a new strategy rule that will maximize his/her profits. To this end, each agent is modeled with LCS learning abilities and its behavior is monitored. Changes in the behavior are caused by the transition from a uniform pricing system to a system where each producer is paid at his/her supply price. Additionally, the possibility of cooperation between two agents when the rest Producers make offers in their cost limit is also observed.

Finally, Petrov and Sheble [17] introduced Genetic Programming in their simulation and tried to model the bipartite Auctions for the Electric Energy Market, by the use of agents. One of the players incorporates knowledge represented as a Decision-Making Tree, which is developed by Genetic Programming. The rest of the agent-players incorporate ruled-defined behaviors.

Special mention should be made to the work of Rosenschein and Zlotkin who, in the early 90s, laid the foundations of the Multi-Agent Negotiation Systems [18]. Within their work, they set the relatively important attribute standards that someone should take into consideration. These are: a) Efficiency (with respect to Pareto and Global Optimality), b) Stability, c) Simplicity (as low computational demands as possible, thus increasing stability and efficiency), d) Distribution and e) Symmetry. In respect with the previously mentioned related work, the Multi-Agent platform implemented in our work tries to take it to the next level. While the implementations mentioned are solutions to specific problems and most of them utilize a preassigned DM technique, "Cassandra" is built in such way that can be easily "transformed" into a prediction tool for every occasion. New prediction models can be easily created and substitute the previously used ones, the time windows used for prediction can change (from day, to week, to month, to year) and also we can combine prediction models with time windows to take more accurate and probable results.

3 Developed System

Cassandra is a multi-agent platform designed and developed to function in an automatic and semi-autonomous manner in Energy Markets. Cassandra employs DM techniques in order to forecast the Settlement Prices, as well as the power load of the Day-Ahead Market. The implemented system may even function in a fully autonomous manner if granted permission, and may proceed with the necessary actions for the establishing a contract. The efficiency of the system is highly dependent on the models generated from historical data, as well as a set of the fail-safe rule base specified by the Power Market expert. Through Cassandra's interface, DM models are re-built dynamically, in order to study, evaluate and compare the results and choose the one that optimally projects running market trends.

3.1 Cassandra Architecture

Cassandra follows the IRF Architecture Model (Intelligent Recommendation Framework) [19], which defines a 4-layer functional model for the agent system. IRF is usually employed in enterprises for the optimization of the administration mechanism, since it can automate and integrate all the data producing or data demanding facets of a company.

Taking a closer look of the Power Market through the IRF prism, one may identify several tasks that have to be tackled:

- Collection of the historical data from previous auctions and processing.
- Application of the suitable DM algorithms in order to build the necessary forecasting models.
- Integration of generated models in the Business Intelligence of the System and evaluation of the results.
- Continuous monitoring Stock Market.

As expected, Cassandra employs a modular architecture (Figure 2), where each module is responsible for one of the aforementioned tasks. The platform also provides a wrapper around all modules and ensures communication with the system users. The modules comprising Cassandra are:

1. **Data Collection Module (DCM):** It is responsible for the collection of historical data, either from files provided by the user, or directly from the Internet.
2. **Data Processing and Mining Module (DPMM):** One of the core modules of Cassandra, which is responsible for the processing of the data, preparing the training sets and applying the DM algorithms.
3. **Decision Making Module (DMM):** It aggregates all information in order to make the optimal decision in any given occasion.
4. **Graphic User Interface Module (GUIM):** It interacts with the users of the System. It must be user-friendly, and easily comprehensive.

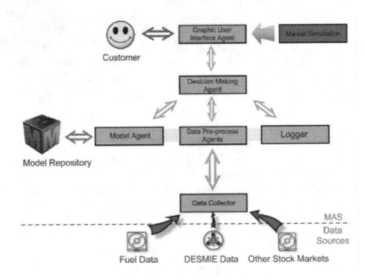

Fig. 2. The Cassandra's 4-layer Architecture

3.2   Cassandra Users

Cassandra identifies three types of users:

– System Analyst
The System Analyst (SA) is an expert user. He/she is the creator of the system, so he/she has absolute knowledge over it. SA is responsible for the smooth operation (unimpeded operation of the system, satisfaction of the software requirements of the users), as well as its efficiency. SA also provides domain knowledge, while he/she is responsible for generating the DM models, in order to decide on the best-performing one(s). Finally, SA checks the decision routines, in order to trace in time any errors that may occur.

– System Manager
The System Manager (SM) is also an expert user, not on agents and DM, but in the Power Stock Market. SM cannot interfere directly with the system, but using his/her experience in the Market SM can easily evaluate the strategic moves of the system and decide whether Cassandra operates efficiently enough or not. SM has the authority to manually override the prediction system (for example change the bids on the Settlement Price), but he/she cannot change the models used by the system for prediction (that is under the Analyst's jurisdiction).

– Monitoring User
The Monitoring User (MU) is the na?ve user of the system. MU only monitors the Power Market moves and the logs the decisions - Cassandra's placed bids. In case MU notices something 'out of the ordinary' (actions that do not have the expected results), MU notifies the SM. SM must then double check MU's observations and, in case of error, notify the SA, who will react accordingly in order to optimize system operation.

Each user group is accommodated through different views of the system, enabled upon user authentication. Table 1 summarizes the functionality provided to each user group.

Table 1. Cassandra Functionality

User Type	Views	Privileges	Functionality
MU	– Monitoring	None	– Account Checking
			– Market Monitoring
			– www/RSS Browsing
			– Chart Viewing
SM	– Monitoring	Managing only	All the above plus:
	– Manager		– Manual Bidding
			– Logger Handling
SA	– Monitoring	All	All the above plus:
	– Manager		– Agent Overview
	– Analyst		– Model Retraining
			– Model Configuration
			– Model Statistics
			– Cost Evaluator

3.3   Implementation

Cassandra provides a multi-functional user interface to facilitate its usage. It has been implemented in Java 1.6 and all agents are developed over the Java Agent Development Framework (JADE) [20], which conforms to the FIPA specifications [21]. The system provides authorized users (SAs) full control over the agents, from their creation till their termination. All DM models are built and evaluated on the WEKA (Waikato Environment for Knowledge Analysis) API [22].

As already denoted, Cassandra supports MUs, SMs, and SAs through respective views. When initialized, Cassandra projects the MU View (Figure 3a). The user has then the option to log in as an SM or an SA (through the 'Settings' menu) and enjoy the privileges of each user category. The SM View (Figure 3b) provides many additional features in comparison to the MU View. It realizes the Logger, a mechanism that

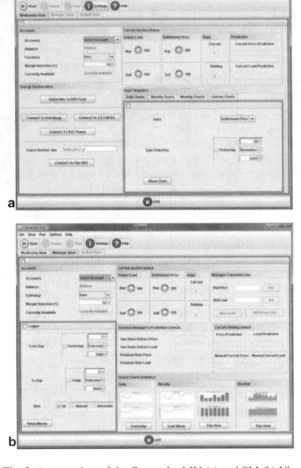

Fig. 3. An overview of the Cassandra MU (a) and SM (b) View

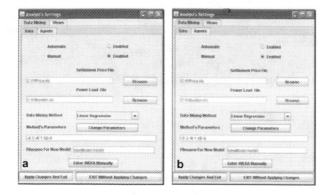

Fig. 4. The Agent and Data Mining Configuration panes

provides a detailed overview of the bidding agent's predictions and actions, in order to monitor and evaluate the efficiency of the system. It also provides an override mechanism for manually bypassing the automated bidding process, in case SM considers the prediction to be faulty. On the other hand, the SA View comprises several useful design and development tools. Such tools are the Cost Evaluator, which calculates the cost of Production given the right Parameters, as well as a Agent and a Data Mining Configuration pane (Figure 4a and 4b, respectively), that provide full control on the MAS architecture (change the types and number of agents residing in the Processing and DM Layer, select different preprocessing agent behaviors) and the creation and evaluation of DM models (select new training and testing datasets, new algorithms/algorithm parameters etc), respectively.

4 Preliminary Results

Before building the Cassandra system, we performed a thorough analysis on available data, in order to build DM models that would efficiently predict the settlement price and the power load in a Day-ahead market. Both problems were modeled as regression problems, where the desired output would be the Decision Manager's prediction of the Settlement Price or Power Load with respect to past auction data. Various experiments and models were built, taking hourly/weekly/monthly periodicity into account. For the shake of simplicity, we provide the results on the models created based on a daily time window (the prior 23 hours-auctions are considered as input, requesting to predict the Settlement Prices and Power Loads for the 24th hour). It should be mentioned that we choose one of the possible prediction models in order to show "Cassandra's" functionality. There are several other approaches such as time windows (taking into consideration the last week's, month's or year's values) as well as combinations of time windows and daily prediction models. Additionally, there may be many data sources to take into consideration such as fuel prices, other Stock Markets etc. All these parameters and methodologies will be further analyzed on the next level of our research.

In this point, we should underline that our system's stability is based on Nash Equilibrium [23]. We are able to change our architecture dynamically so to follow the changes of the market, with respect to the other players strategies, but always trying to reach to the Equilibrium Point. As far as the efficiency of our implementation is concerned, the DM techniques applied to produce the prediction models, as shown by the next chapters, are giving very promising results. If these are combined with more complex techniques as time windows will make our platform very efficient.

The WEKA suite was employed for the conduction of the experiments with a plethora of algorithms over the available datasets.

4.1 Training

Five different classification (regression) schemes were applied, in order to decide on the one that optimally meets the problem of predicting the Settlement Price and the Power Load for the 24th hour: a) Simple Linear Regression, b) Linear Regression, c) Isotonic Regression, d) Pace Regression, and e) Additive Regression [24]. The Simple Linear Regression models applied, as expected, gave poor results on every dataset it was applied on, since it is known to a very provincial technique. The correlation coefficiency (cc) of the model extracted was around 0.95 and 0.98, while the Relative Absolute Error (RAE) was around 26% and 22% for the Settlement Price and the Power Load, respectively. A large number of experiments were conducted applying Linear Regression, on different datasets (varying size) and with different algorithm parameters. The cc was a slightly better (0.96 and 0.98-1), as well as the RAE for the Settlement Price

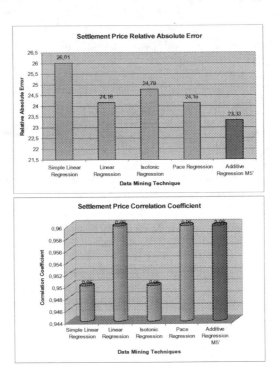

Fig. 5. A comparison of the regression schemes applied on the Settlement price

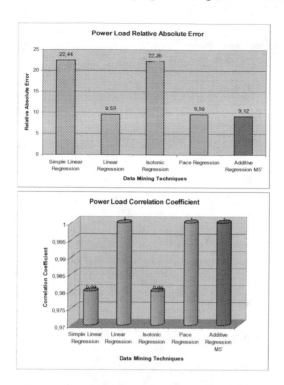

Fig. 6. A comparison of the regression schemes applied on the Settlement price

(24%). The RAE for the Power Load (10%), though, improved significantly. With the application of Isotonic Regression techniques, the results were as disappointing as in the case of Simple Linear Regression, only with a slight improvement in RAE for the Settlement Price (24%). The Pace Regression algorithm (numerous parameters for algorithm fine-tuning), came up with very good results both for cc (0.96, 1) and for RAE (24%, 10%) for both goods. It's should be mentioned that increasing or decreasing the Estimator's parameters didn't significantly influence the resulting efficiency, while in some cases the Mean Squared Error was much higher that the simple techniques used, indicating overfitting.

4.2 Meta Classification

Apart from the regression schemas applied, we also tested some meta-classifier schemas, striving for optimal performance. Various Additive Regression schemes were tested against the same datasets, in order to ensure equally compared test results.

Three schemes applied to the dataset: a) *DecisionStamp* (building and using decision stamp), *REPTree* (fast decision tree learner) and *M5'* classifiers. The first performed worse than any other model extracted (*cc* 0.93 and 0.96, *RAE* 33% and 31%, respectively). Nevertheless, the *REPTree* schema came up with much better results (*cc* was

0.95 and 0.99, while *RAE* improved to 23.21% and 14.93%, respectively). Finally, *M5'* outperformed all schemes and algorithms, with cc reaching the maximum value of 0.96 and 1, while in the same time *RAE* was around 23% and 9%, respectively.

4.3 Data Mining Experiment Results

It's obvious that the combination of the Additive Regression meta-classification schema with the M5' regression algorithm, applied on the given datasets significantly outperforms all the other learning methods applied. The main advantage of M5' with respect to the other methods applied is that it produced a simple and compact tree model, in contrast with Simple and plain Linear Regression that attempted to impose a linear relationship on the data, Isotonic Regression that picks the attribute that results the lowest square error, and Pace Regression that is optimal when the number of coefficients of the linear model tends to infinity. Combined with Additive Regression, which introduces a stochastic factor, *M5'* succeeded in improving the accuracy of the predictors and achieved optimal performance. Figures 5 and 6 provide a qualitative comparison between the regression schemes applied, cross-validated on the same dataset.

5   Pilot Case Scenario

In order to demonstrate the proper functioning of Cassandra, we have set up a real-life scenario, based on historical data for the Greek Energy Stock Market[1].

Within this Day-ahead market, Producers place each day 24-bid bundles for the forthcoming day's 24-hourly auctions, while Consumers declare their needs in Power Supply, along with the maximum price willing to pay. The ISA takes all this information in consideration and calculates the aggregate Supply and Demand Curves, thus defining the Settlement Price and Power Load for each one of the forthcoming day's auctions. All Producers that have made bids, less or equal than the specified Settlement Price are included in the next day's distribution network, and are paid at the Settlement Price for each MWh sold. The rest of the Producers are not included in the transaction.

The scenario Cassandra is tested against is the following: First, we simulate a power stock market by randomly select 5 days from the historical data on previous auctions. For each day, Cassandra's agents try to predict the 24th-hour auction Settlement Price and Power Load, based on the values of the other 23-hour auctions. Then the predicted values are compared to the actual ones residing in the dataset. In case the predicted price is equal or less than the actual price, we consider bidding to be successful, (Within Market ranges) for that hourly auction. If not, we consider it unsuccessful. Three experiments were conducted, where Cassandra agents employed three different DM models:

1. The first DM model was extracted by the use of Simple Linear regression on raw data (no pre-processing was performed). Figure 7a depicts the results, where one may notice that most predictions were 'Unsuccessful'.

---

[1] http://www.desmie.gr/

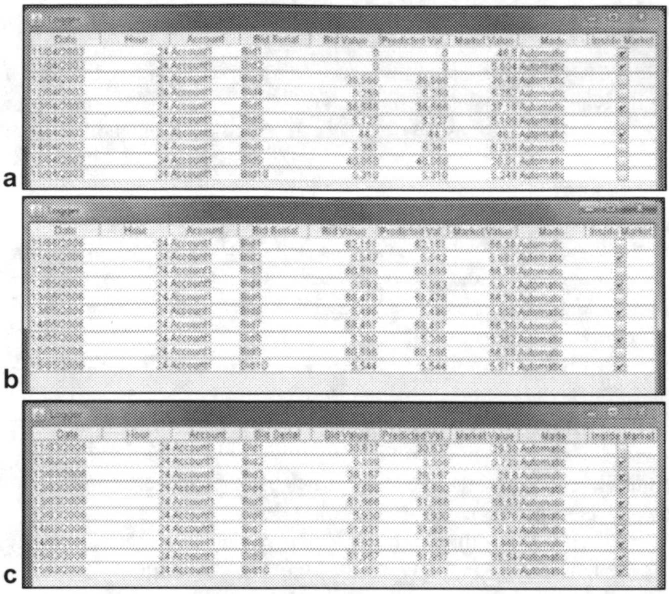

Fig. 7. Illustrating the accuracy of the three applied models

2. The second DM model was again extracted by the use of Simple Linear regression, this time on a filtered (preprocessed) dataset. Figure 7b depicts the results, where improvement can be seen.
3. Finally, the third DM model was extracted by the use of the Additive Regression with the M5' regression scheme. Figure 7c depicts the results, where improvement is obvious. 3.

6 Future Work

Within the context of this paper we have presented Cassandra, a multi-agent system that employs DM primitives in order to automate the process of participating in the Power Stock Market. Cassandra succeeds in predicting forthcoming Settlement Prices and Power Load values, allowing its 'master' to bid within market ranges and maximize profit. The tool provides the ability to design the MAS and decide on the technique and algorithm on which to build the DM model on, while it provides a number of utilities for monitoring the market and analyzing trends.

Future work is focused on two directions: a) extensively study periodicity (same day each month, same weekend each year and so on), in order to identify an even more efficient model for staying 'within market ranges' and b) identify the maximum price to bid so as to maximize revenue. Additionally, one may work towards improving the GUI (Graphical User Interface) of the platform. The addition of more graphs, tables and diagrams would help in extracting useful information produced during the system's operation.

References

1. Amin, M.: Restructuring the Electric Enterprise. In: Market Analysis and Resource Management, pp. 2–16. Kluwer Publishers, Dordrecht (2002)
2. Amin, M., Ballard, D.: Defining new markets for intelligent agents. IT Professional 2(4), 29–35 (2000)
3. Bagnall, A.J., Smith, G.D.: Game playing with autonomous adaptive agents in a simplified economic model of the uk market in electricity generation. In: IEEE-PES / CSEE International Conference on Power System Technology, POWERCON 2000, pp. 891–896 (2000)
4. Bellifemine, F., Poggi, A., Rimassa, R.: Developing multi-agent systems with JADE. In: Castelfranchi, C., Lespérance, Y. (eds.) ATAL 2000. LNCS, vol. 1986, pp. 89–101. Springer, Heidelberg (2001)
5. Conzelmann, G., Boyd, G., Koritarov, V., Veselka, T.: Multi-agent power market simulation using emcas 3, 2829–2834 (2005)
6. Freedman, D.: Statistical Models: Theory and Practice. Cambridge University Press, Cambridge (2005)
7. He, M., Jennings, N.R., fung Leung, H.: On agent-mediated electronic commerce. IEEE Trans. Knowl. Data Eng. 15(4), 985–1003 (2003)
8. Holland, J.H.: Genetic Algorithms and Classifier Systems: Foundations and Future Directions. In: Grefenstette, J.J. (ed.) Proceedings of the 2nd International Conference on Genetic Algorithms (ICGA 1987), Cambridge, MA, pp. 82–89. Lawrence Erlbaum Associates (1987)
9. Koesrindartoto, D.P., Sun, J.: An agent-based computational laboratory for testing the economic reliability of wholesale power market designs. Computing in Economics and Finance 2005 50, Society for Computational Economics (November 2005)
10. Koesrindartoto, D.P., Tesfatsion, L.S.: Testing the reliability of ferc's wholesale power market platform: An agent-based computational economics approach. Staff General Research Papers 12326, Iowa State University, Department of Economics (May 2005)
11. Petrov, V., Sheble, G.: Power auctions bid generation with adaptive agents using genetic programming. In: I. of Electrical and E. Engineers (eds.) Proceedings of the 2000 North American Power Symposium (2000)
12. Praca, I., Ramos, C., Vale, Z., Cordeiro, M.: Mascem: a multiagent system that simulates competitive electricity markets. IEEE Intelligent Systems 18(6), 54–60 (2003)
13. Praca, I., Ramos, C., Vale, Z., Cordeiro, M.: Intelligent agents for negotiation and game-based decision support in electricity markets. Engineering intelligent systems for electrical engineering and communications 13(2), 147–154 (2005)
14. Praca, I., Ramos, C., Vale, Z., Cordeiro, M.: Testing the scenario analysis algorithm of an agent-based simulator for competitive electricity markets. In: ECMS (ed.) Proceedings 19th European Conference on Modeling and Simulation (2005)
15. Rosenschein, J.S., Zlotkin, G.: Rules of Encounter: Designing Conventions for Automated Negotiation Among Computers. MIT Press, Cambridge (1994)
16. Stone, P.: Learning and multiagent reasoning for autonomous agents. In: 20th International Joint Conference on Artificial Intelligence, pp. 13–30 (2007)
17. Symeonidis, A.L., Mitkas, P.A.: Agent Intelligence Through Data Mining. Springer Science and Business Media (2005)
18. Tellidou, A., Bakirtzis, A.: Multi-agent reinforcement learning for strategic bidding in power markets, pp. 408–413 (September 2006)
19. The FIPA Foundations. Foundation for intelligent physical agents specifications. Technical report, The FIPA Consortium (2003)
20. Vickrey, W.: Counterspeculation, auctions, and competitive sealed tenders. The Journal of Finance 16(1), 8–37 (1961)

21. Watkins, C.J.C.H.: Learning from Delayed Rewards. PhD thesis, King's College, Cambridge, UK (1989)
22. Witten, I.H., Frank, E.: Data Mining: Practical Machine Learning Tools and Techniques with Java Implementations. Morgan Kaufman, San Francisco (2000)
23. Wurman, P.R., Wellman, M.P., Walsh, W.E.: The michigan internet auctionbot: a configurable auction server for human and software agents. In: Second International Conference on Autonomous Agents, pp. 301–308. ACM Press, New York (1998)

Enhancing Agent Intelligence through Data Mining: A Power Plant Case Study

Christina Athanasopoulou and Vasilis Chatziathanasiou

Department of Electrical and Computer Engineering,
Aristotle University of Thessaloniki, Thessaloniki, Greece
{athanasc,hatziath}@eng.auth.gr

Abstract. In this paper, the methodology for an intelligent assistant for power plants is presented. Multiagent systems technology and data mining techniques are combined to enhance the intelligence of the proposed application, mainly in two aspects: increase the reliability of input data (sensor validation and false measurement replacement) and generate new control monitoring rules. Various classification algorithms are compared. The performance of the application, as tested via simulation experiments, is discussed.

Keywords: Intelligent agents, data mining, power plant.

1   Introduction

The deregulation of the electric power and the continuously stricter environmental terms, imposed by international protocols, call for more efficient and more economic Thermal Power Plant (TPP) operation. In order to achieve optimum plant operation, electric companies embrace new, advanced control monitoring systems. The confrontation of the complexity of these new technologies and the profitable exploitation of the plethora of data produced by them necessitate the provision of intelligent tools that will assist the personnel to detect and to face the operational problems in real time.

In this direction companies as ABB and Metso renew their commercial applications mainly by adding tools that supervise the trends of operational parameters [1,2]. At the same time, research groups investigate the possibility of applying artificial intelligence methods for the production of intelligent tools that will solve TPP operating problems [3,4,5].

However, the existing solutions in the literature usually are designated for specific plant technology (eg. boiler Benson or Sultzer, with/without low-NOx burners etc), rendering thus the adaptation expensive and time-consuming, if not impossible. Besides, most of the proposed methods are developed for the confrontation of particular problems and are not applicable to other cases [4,5].

What is missing is a generalized framework for the design and development of intelligent applications for the support of the TPP operation control. Our research work introduces a framework for the formation of an adaptable and extendable system, that can correspond to all the TPP needs from the reduction of gas emissions to the timely

L. Cao et al. (Eds.): ADMI 2009, LNCS 5680, pp. 126–138, 2009.

breakdown avoidance. It combines latest advances in three research areas; knowledge engineering (KE), data mining (DM) and multi-agent systems (MASs). The strength of the proposed framework lies in the embedment of DM models into agents.

The rest of this paper is organized as follows. Section 2 discusses the main issues that the proposed application faces. The three research areas, which form the basis of our approach, are listed in section 3. Section 4 introduces the proposed methodology. The MAS architecture is outlined in section 5. Issues concerning the extraction of DM models are discussed in section 6, while the performance of the DM algorithms in section 7. Section 8 includes some evaluation points resulting from simulation experiments and section 9 the conclusions.

2   Main IPPAMAS Issues

The proposed Intelligent Power Plant engineer Assistant Multi-Agent System (IPPA-MAS) is an integrated tool that extends from the sensor data validation to the provision of information and indications to the power plant personnel. It initially evaluates and improves the reliability of the plant measurements in order to reassure that data used by the next stages are correct. Then it exploits the historical data by applying data mining techniques and combines the derived DM models with the current information, so as to produce knowledge and provide the adequate indications to the plant personnel. Finally, it diffuses the produced information to the appropriate users, in the suitable form, in the right time and place. Selective literature for these three main issues, along with the proposed approach, are given in the following paragraphs.

2.1   Sensor Validation and Estimation

The evaluation of information on which the decision-making for an action is based, plays a particularly important role in the industry. Erroneous data can lead to critical situations because of wrong handlings. The values of the sensors, as thermometers and manometers, are considered as information in the case of power plants.

Typical approaches for the detection of incorrect readings of a sensor include the use of hardware redundancy and majority voting, analytical redundancy, and temporal redundancy [6]. Artificial Intelligence techniques have also been proposed for sensor validation and sensor values estimation [3, 7, 8]. Nevertheless, these various methods have serious drawbacks in practice (increased demands in human expertise and computer resources, time consuming etc).

At the current project it was decided to base the validation on rules that derive from the measuring equipment requirements, the plant operation specifications and the personnel experience [9]. As far as sensor values estimation is concerned, DM algorithms are applied for deriving models that estimate the value of one variable based on others used as input parameters. The estimated values can then be used instead of the ones recorded by a measuring instrument out of order. Software agents apply both the validation rules and the DM models during the on-line monitoring procedure.

2.2   Decision Support

The potential contribution of intelligent MAS to decision support systems (DSSs) is appreciated as significant by various researchers [10, 11]. The MAS technology has been adopted in many research projects for supporting the users at the decision-making procedure [11, 12].

The main part of the presented MAS evaluates the current conditions on-line and provides the users with the appropriate indications. Also, it locates the cases that the installed commercial control monitoring system fails to produce the proper alarm signals and undertakes their activation (eg. when the measurement of a variable associated with an alarm is erroneous).

2.3 Information Flow

The right information management and distribution plays a key role for an enterprise. MAS technology has been used for this aim in many applications in sectors as the power industry and logistics [13, 14]. A MAS for information management is able to incorporate and handle the heterogeneous, distributed sources of information in a TPP and to model the various users. Main advantages of the proposed system, as far as information fusion is concerned, are the ubiquity and the context-awareness, which are introduced [15].

3 Research Areas

3.1 Knowledge Engineering

Domain knowledge is considered to be one of the determinants in a large-scale DM project [16]. It is important to mention that knowledge on the plant operation is not explicitly expressed by the persons who possess it, neither it is easily explained and standardized in books or handbooks. Knowledge engineering provided the scientific way of transforming the implicit knowledge of personnel into explicit, so as to exploit it for the development of the proposed application. In addition, it supplied the tools for modeling the problem and producing specifications comprehensible, hence easily applicable.

3.2 Data Mining

The Knowledge Discovery in Databases (KDD) procedure was applied for the correlation of operational parameters for which there was no known relation to connect them (i.e. from thermodynamics, plant specifications etc.). Its advantages comparatively to conventional statistical methods are that it offers techniques for confronting the vast volume of historical plant operational data and that it does not prerequisite the determination of specific questions.

3.3 Multiagent System

The control system of a TPP must meet increasingly demanding requirements stemming from the need to cope with significant degrees of uncertainty. MASs have recently emerged as a powerful technology to face the complexity of a power plant. Indeed, several experiences already testify to the advantages of using agents in control monitoring [17, 18]. The autonomy of MAS components reflects the intrinsically decentralised nature of modern distributed control monitoring systems. The flexible way in which agents operate and interact (both with each other and with the environment) is apt to the dynamic and unpredictable TPP operation. In addition, agents are most appropriate for developing context-aware and ubiquitous applications [15, 19], that will offer their services all over the plant premises and will be able to adapt to different situations, from normal operation to a crisis resolution. Finally, agents are most suitable for embedding DM models [20]. This was significantly important for the presented application, as its intelligence is mainly based on these models.

4   Methodology

For the design and development of the system, KE was chosen for modelling the problem, KDD for handling the enormous volume of operation data and facing the lack of known equation connecting them, and MAS for the confrontation of the complexity and the distributed nature of the system.

It was argued that DM techniques were adopted for enhancing the MAS intelligence by extracting knowledge from the vast amounts of operation data. However, a critical sector, as the power industry, cannot unquestioningly rely on automatically extracted statistical models, as there is the risk to come up with good statistical results that have no physical meaning. Such models would not reflect the actual operation, thus could lead to dangerous choices, if applied. In order to eliminate this risk, a KE method was utilized to capture domain knowledge for evaluating the appropriateness of the DM models. Furthermore, through KE the bounds of using DM models were set: prerequisites of a certain action consist one boundary and the conditions that necessitates it, the other one (Fig.1). The DM models were used to suggest on cases that lie in between. The MAS was designed appropriately to apply the rules and the models on-line.

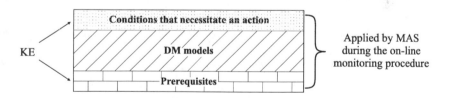

Fig. 1. Data mining application area

Our proposed methodology consists of the following five steps:

1. Concentration and formulation of information and implicit knowledge possessed by the plant personnel. Application of a KE method for modelling the plant and specifying the IPPAMAS thoroughly.
2. Application of the KDD on the plant operation historical data for extracting the most suitable models. The models are used for estimating the values of variables and for modelling actions.
3. Embedment of derived models in agents for the replacement of erroneous sensor measurements and for the calculation of optimum value of operation parameters.

4. Development of the MAS for the management of data, the application of models and rules and the configuration of appropriate indications to the TPP personnel.
5. Deployment of a Wireless Local Area Network (WLAN) combined with a (pre-existing or not) wired Ethernet LAN for the diffusion of the services allover the plant premises.

A KE methodology, CommonKADS , was chosen for the case of a TPP [21]. It offers a predefined set of models that together provide a comprehensive view of the project under development. The decision to base the development of the system on MAS predicated the adoption of its extension MASCommonKADS , which adds a model capturing particular aspects of the agents [22].

For the application of the data preprocessing techniques and the classification algorithms, the widely used Waikato Environment for Knowledge Analysis (WEKA) was chosen [23, 24].

The Java Agent Development framework (JADE) was used for the implementation of the MAS [25]. Particularly, for deploying the Data-Mining Agents the Agent Academy platform (version "Reloaded") was used [26]. A unique functionality offered by Agent Academy is the embedding of models that derive from the application of most of the classification algorithms available in WEKA. As it is implemented upon the JADE infrastructure, it is self-evident that the deployed agents are compatible with the remaining ones created with JADE.

5 MAS Architecture

The MAS architecture is structured in three layers: Sensors Layer, Condition Monitoring Layer and Engineer Assistant Layer. At each layer agents act in groups that correspond to different plant subsystems. There are also central agents that assure the consistency of solutions and the resolution of conflicts.

The Sensor Layer (1st) is responsible for the identification and reconstruction of sensor faults. It ensures that data entered to the application are valid (Fig.2).

The Condition Monitoring Layer (2nd) is responsible for the safe operation of the TPP and its optimization. In this phase, meaning is assigned to data to produce the appropriate information, as alarm signals and suggestions on handlings. Its main functionalities are depicted in Fig.3. The activities that follow the 'Request Estimation' are as shown in Fig.2

The Engineer Assistant Layer (3rd) distributes the information to the appropriate users. Information pieces are handled differently depending on the current operating conditions and the context of the users.

At each layer several agents cooperate for confronting the respective tasks. These agents extend one of the six basic agent types:

1. Variable Agent: identification of faulty sensors
2. Data-Mining Agent: DM models application
3. Recommender Agent: recommendations on actions
4. Condition Agent: alarms triggering

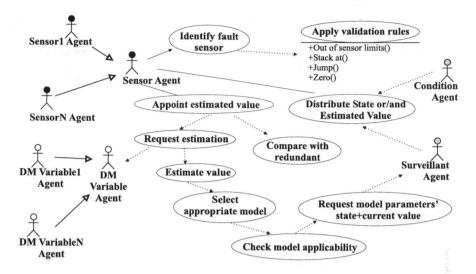

Fig. 2. Sensor validation and fault values replacement with estimated ones

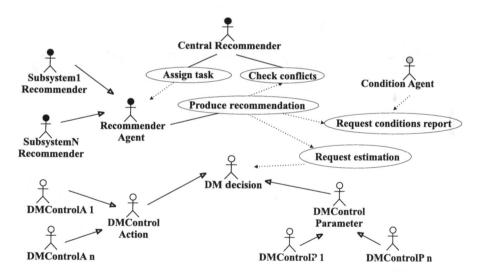

Fig. 3. Produce recommendations on control actions or parameters' regulation

5. Distribution Agent: information distribution
6. User-Interaction Agent: information personalization

There are also four auxiliary agents that provide their services to all levels:

1. Supervisor Agent: identifies the operation mode (so as the other agents apply the appropriate models, rules etc).

2. Sensor Surveillant Agent: gathers and distributes information on sensor measurements.
3. Synchronization Agent: ensures the synchronization of all the agents to the application cycle.
4. DBApplication Agent: handles the data storage and retrieval to and from the database of the application.

In addition, Monitoring Agent is responsible for the MAS coordination. Final, Trainer Agent supervises the MAS retraining procedure, which ensures that the DM models and the rules that are embedded into the agents are up-to-date, thus reflect the current TPP operation.

6 DM Models

DM models are embedded into DM agents that participate at the first two layers, as these concern the plant operation data. The extraction of models concerning the user profiles for the 3rd layer was also investigated, but rejected, as there is not significant space for improvement. The latter is mainly due to two reasons: 1) it is known from the beginning who the users will be (i.e. categories, particularities, skills, preferences etc) and 2) the security specifications of a TPP restrict the degree of personalization allowed.

For each discrete case more than one DM model is extracted and stored at the application's repository. The respective DM agent chooses the most appropriate of them when requested on-line. These models derive from:

1. Applying different classification algorithms to the same dataset. This is done in order to have models selected with different criteria, as statistical metrics and existence of missing data or outliers. Depending on the current circumstances one of them might be more appropriate than the others.
2. Applying the same classification algorithm (the one presenting the best performance) to different datasets. These datasets are selected from the initial dataset by applying attribute selection methods. Having models with different input parameters offer the flexibility to apply each time the one whose parameters are all valid at the current moment.

At the Sensor Layer, one agent that extends the Sensor Agent type is created for each sensor. Obviously, this does not concern all the plant sensors; a part of them was chosen based on their significance for the plant operation, linkage to alarm signals and usage by the application itself. In the event of a false reading, an estimated value is used to replace the recorded value of the measurement instrument. The task of estimating a value is undertaken by an agent that extends the basic Data-Mining Agent. As depicted in Fig.2, the agent 'Selects the appropriate model' among the available ones and 'Checks the model applicability'.

At the Condition Monitoring Layer, agents that extend the basic types DMControlAction and DMControlParameter apply DM models for advising on actions and for calculating the optimum value of controllable parameters (in order to tune a subsystem appropriately). Likewise at the 1st layer, two or more models for each examined case are stored to the application's repository.

The DM models used in the Condition Monitoring Layer were not derived from the whole data. The dataset comprised of only a part of the historical data that reflected the best system operation judged according to selected criteria. These criteria varies depending on the goals set for each subsystem. For instance, the criterion for the Flue-gas subsystem is the reduction of one or more of the flue-gas emissions (NOx, CO_2 etc.).

7 DM Results

The KDD was followed for deriving the DM models. A detailed description of the stages proceeding DM is beyond the scope of this paper (see [9]).

Data were taken from the TPP Meliti of the Public Power Corporation, Greece. They concerned three plant subsystems. For each subsystem 80-100 variables were selected. The records covered one year of almost continuous operation. The datasets were split according to the operation mode (eg. full/low load). The one concerning full plant operation represented 70% of the data, approximately. It contained, after the cleaning, more than 360000 instances. These were unequally distributed to 15 files of different length.

The example presented in this paper concerns the Flue-gas subsystem of the boiler. The original dataset for this specific subsystem comprised of 88 variables (temperatures, pressures, flows, etc.), which were reduced to 44 by applying aggregation techniques. In an effort to further reduce the input parameters, attribute selection filters were tested [9].

Initially, we experimented with a range of classifiers, diversifying their parameters and the combinations with pre-processing techniques. Table 1 includes three statistic metrics for the performance of 21 classifiers (implementation was provided by WEKA), when applied to a dataset comprised of 10252 rows of data for modeling the variable 'Flue-gas temperature after the Economiser' (i.e. at the exit of the boiler).

For comparison reasons the statistics listed were the best ones that resulted from the application of each algorithm after modifying the initial parameters values set in WEKA. Some of them performed significantly better (or worse) with modified parameters, while the performance of others was not affected considerably by this. It should be noted that the performance of the meta-algorithms was judged in comparison with the performance of the base algorithm on top of which they were applied.

Five classification algorithms were selected for the remaining of the experiments based on their statistics and the response time. The REPTree gave satisfactory results in the majority of the cases (Fig.4a). It builds a regression tree using information gain [24]. It had minimum execution times, so it was used for evaluating the selected datasets and the appropriateness of the various preprocessing techniques.

Next in line came algorithms that had good performance, but were time consuming. M5Rules generates a decision list for regression problems using separate-and-conquer (it is based on the M5 algorithm) [27]. Its output is most suitable for communicating the dm models to the plant experts, as it comprises of rules easily understandable by humans. Its performance was among the best ones in most of the experiments (Fig.4b).

MultilayerPerceptron algorithm, as its name indicates, is a neural network (NN) which uses backpropagation to train [28]. NN are quite often applied by various researchers

Table 1. Statistical results of various classification algorithms applied for the modeling of the variable 'Flue gas temperature after the Economiser'. Mean value= 294.5 °C.

	Algorithm	Correlation Coefficient	Mean Absolute Error(°C)	Relative Absolute Error(%)
Functions	Linear Regression	0.812	1.845	60.19
	Least Med Sq	0.560	2.068	67.47
	MultilayerPerceptron	0.994	0.363	11.86
	Simple Linear Regression	0.630	2.330	75.83
	RBF Network	0.786	1.65	53.95
	SMOreg	0.772	1.649	53.79
	Pace regression	0.772	1.642	53.65
Lazy	IB1	0.987	0.167	5.45
	IBk (k=2)	0.988	0.166	5.42
	Kstar	0.998	0.017	0.04
	LWL	0.732	2.111	68.86
Rule	Conjuctive Rule	0.784	2.016	65.77
	Decision Table	0.993	0.173	5.64
	M5Rules	0.992	0.226	7.38
Tree	Decision Stump	0.698	2.225	72.57
	M5P	0.997	0.171	5.59
	REPTree	0.992	0.170	5.55
Meta	Additive Regression (Decision Stump)	0.882	1.441	47.00
	Bagging (Decision Stump)	0.699	2.221	72.43
	Regression By Discretization (J48)	0.993	0.229	7.48
	CVParameterSelection (REPTree)	0.994	0.165	5.54

in papers relative to our work regarding the application domain (eg. for sensor validation/estimation), so MultilayerPerceptron was mainly chosen for comparison reasons. Its performance was average (Fig.4c).

Finally, the algorithms IBk (k=2) and Kstar (K*) usually demonstrated the best performance, as indicatively is shown in Fig.4e-4f and Fig.4d, respectively. They both belong to the category of instance-based classifiers. IBk is a k-nearest neighbours classifier that normalizes attributes by default [29]. K* differs from other instance-based learners in that it uses an entropy-based distance function [30].

However the requirements in computer resources of the last four ones discouraged or even stopped their application, depending on the dataset dimensions. For these reasons it was decided to apply the REPTree at the beginning for discarding the datasets and then apply all five of them for extracting the final models. As aforementioned, we decided to use different algorithms for extracting the DM models because each of them handles effectively different cases, i.e. missing data, outliers etc. Consequently,

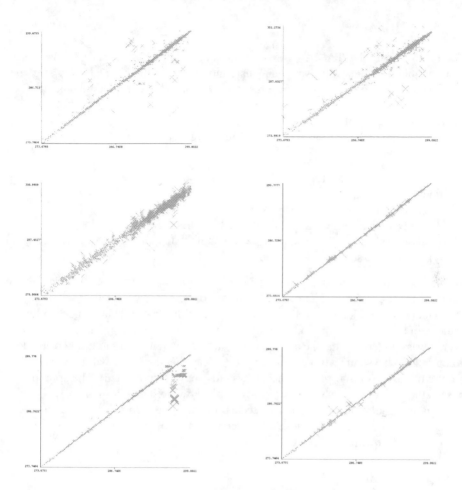

Fig. 4. The performance of five classification algorithms applied for modelling the variable Flue gas Temperature after the Economiser: a-REPTree, M5Rules, c-MultilayerPerceptron, d-K*, e-IBk, f-IBk (no missing values). The axes are as follows: x: measured values and y: predicted values.

depending on the case, there is always a proper model available and the whole system responds effectively. The above-mentioned are equally important for the retraining procedure. Indicatively, the performance of some algorithms, including IBk, is affected by the presence of missing data (Fig.4e-Fig.4f), while others, as REPTree, deal quite efficiently with them.

Apart from the performance metrics, the results were also evaluated based on their physical meaning (correlation of variables). This also involved knowledge interpretation by domain experts. This necessitate adequate knowledge representation formalism (eg. rules defined by a decision tree). As aforementioned, the algorithm M5Rules was usually applied for this purpose.

8 Evaluation

The performance of IPPAMAS was tested through simulation of several cases. Several possible problems were simultaneously introduced in the simulation, as the non-scheduled termination of an agent, presence of false sensor measurements, etc. The MAS deployed for the evaluation comprised of 78 agents and concerned three boiler subsystems. The simulation studies revealed several issues with respect to the service prototype. Three are the points that are directly related with the embedment of DM models in agents:

1. Response time. In all cases the MAS corresponded successfully within the predefined time period of 1 minute.
2. Accuracy. The estimated values were within the acceptable fault tolerance as predefined by the experts (eg. $1°C$ for the Flue gas Temperature after the Economiser).
3. Reliability. The MAS proved that it can handle many simultaneous cases and that it produces the same output for the same input.

With respect to the content, the indications for actions were found appropriate in the majority of cases (70%). The remaining were considered unnecessary or overdue by the plant personnel, leading to the conclusion that either more data are needed for the training of the system or the specifications (predefined rules etc) are not enough. However, it was pointed out that the percentage of unsuccessful indications might be reduced when more subsystems will be monitored by the application, as it will have an overall picture of the operation (most of the subsystems are interconnected). The appropriateness of the DM models, which was already implied by the statistics metrics, was proved through the simulation. The agents succeeded in choosing the most suitable model depending on the circumstances. They also accomplish to offer solutions and to overcome problems by following alternative scenarios.

9 Conclusion

In this paper, a MAS and DM techniques are combined in order to reproduce effectively the complex power plant operation, which is difficult to model otherwise. This combination increases the adaptability and the extensibility of the system; the extraction of new DM models based on new operation data is sufficient for capturing changes that concern the TPP operation (caused by wear, maintenance, replacement, or even addition of new mechanical equipment). The DM models enhance the MAS intelligence so as to successfully replace false sensor measurements and to produce the appropriate indications for actions to the plant personnel. Through the MAS the DM models are exploited so as to function not solely as a tool for calculations, but as a support tool for the personnel.

The short run plans include the design and configuration of more simulation experiments that may reveal more points that call for attention and on the same time give rise to innovative ideas.

Finally, as future work is concerned, an interesting topic for research would be the application of DM techniques to the application data of the MAS, in order to improve the performance of the agents.

Acknowledgments

The research work presented in this paper is partially funded by the 03ED735 research project, implemented within the framework of the "Reinforcement Programme of Human Research Manpower" (PENED) and co-financed by National and Community Funds (25% from the Greek Ministry of Development-General Secretariat of Research and Technology and 75% from E.U.-European Social Funding).

Special thanks are owed to the engineers, technical personnel and operators of the Public Power Corporation S.A. thermal power plants of Western Macedonia, Greece for providing the information, data and continuous support.

References

1. ABB Group, Products & Services, http://www.abb.com/ProductGuide
2. Metso, http://www.metsoautomation.com
3. Flynn, D. (ed.): Thermal Power Plant Simulation and Control. IEE, London (2003)
4. Hadjiski, M., Boshnakov, K., Christova, N., Terziev, A.: Multi Agent Simulation in Inference Evaluation of Steam Boiler Emission. In: 19th European Conference on Modeling and Simulation, pp. 552–557. Riga, Latvia (2005)
5. Ma, Z., Iman, F., Lu, P., Sears, R., Kong, L., Rokanuzzaman, A.S., McCollor, D.P., Benson, S.A.: A Comprehensive Slagging and Fouling Prediction Tool for Coal-Fired Boilers and its Validation/Application. Fuel Process. Technol. 88, 1035–1043 (2007)
6. Frank, P.: Fault Diagnosis in Dynamic Systems Using Analytical and Knowledge Based Redundancy- a Survey and Some New Results. Automatica 26, 459–470 (1990)
7. Eryurek, E., Upadhyaya, B.R.: Sensor Validation for Power Plants Using Adaptive Backpropagation Neural Network. IEEE Trans. Nucl. Science 37, 1040–1047 (1990)
8. Ibarguengoytia, P.H., Vadera, S., Sucar, L.E.: A Probabilistic Model for Information and Sensor Validation. The Computer Journal 49(1), 113–126 (2006)
9. Athanasopoulou, C., Chatziathanasiou, V.: Intelligent System for Identification and Replacement of Faulty Sensor Measurements in Thermal Power Plants (IPPAMAS: Part 1). Expert Systems With Applications 36(5), 8750–8757 (2009)
10. Shim, J.: Past, Present, and Future of Decision Support Technology. Decision Support Systems 33(2), 111–126 (2002)
11. Vahidov, R.: Intermediating User-DSS Interaction with Autonomous Agents. IEEE Trans. on Systems, Man, and Cybernetics 35(6), 964–970 (2005)
12. Gao, S., Xu, D.: Conceptual Modeling and Development of an Intelligent Agent-Assisted Decision Support System for Anti-money Laundering. Expert Systems with Applications 36, 1493–1504 (2009)
13. Lucas, C., Zia, M.A., Shirazi, M.R.A., Alishahi, A.: Development of a Multi-agent Information Management System for Iran Power Industry-A Case Study. In: Power Tech 2001 Proceedings, vol. 3. IEEE, Porto (2001)
14. Pechoucek, M., Marik, V.: Industrial Deployment of Multi-agent Technologies: Review and Selected Case Studies. Auton Agent Multi-Agent Syst. 17, 397–431 (2008)
15. Athanasopoulou, C., Chatziathanasiou, V.: Prototype For Optimizing Power Plant Operation. In: Mangina, E., Carbo, J., Molina, J. (eds.) Agent-based Ubiquitous Computing. Atlantis Press (2009)
16. Kopanas, I., Avouris, N.M., Daskalaki, S.: The Role of Domain Knowledge in a Large Scale Data Mining Project. In: Vlahavas, I.P., Spyropoulos, C.D. (eds.) SETN 2002. LNCS (LNAI), vol. 2308, pp. 288–299. Springer, Heidelberg (2002)

17. Arranz, A., Cruz, A., Sanz-Bobi, M.A., Ruiz, P., Coutino, J.: DADICC: Intelligent System for Anomaly Detection in a Combined Cycle Gas Turbine Plant. Expert Systems with Applications 34, 2267–2277 (2008)
18. Mangina, E.: Application of Intelligent Agents in Power Industry: Promises and Complex Issues. In: Marik, V., Muller, J., Pechoucek, M. (eds.) CEEMAS 2003. LNCS (LNAI), vol. 2691, pp. 564–574. Springer, Heidelberg (2003)
19. Soldatos, J., Pandis, I., Stamatis, K., Polymenakos, L., Crowley, J.: Agent Based Middleware Infrastructure for Autonomous Context-Aware Ubiquitous Computing Services. Computer Communic. 30, 577–591 (2007)
20. Symeonidis, A., Mitkas, P.A.: Agent Intelligence Through Data Mining. Springer, New York (2005)
21. Shreiber, G., Akkermans, H., Anjewierden, A., de Hoog, R., Shadbolt, N., Van de Velde, W., Wielinga, B.: Knowledge Engineering and Management: the CommonKADS methodology. MIT Press, Cambridge (2000)
22. Iglesias, C.A., Garijo, M.: The Agent-Oriented Methodology MASCommonKADS. In: Henderson-Sellers, B., Giorgini, P. (eds.) Agent-Oriented Methodologies, pp. 46–78. IDEA Group Publishing (2005)
23. WEKA, http://www.cs.waikato.ac.nz/~ml/weka/index.html
24. Witten, I., Frank, E.: Data Mining: Practical Machine Learning Tools and Techniques, 2nd edn. Morgan Kaufmann, San Francisco (2005)
25. JADE, http://jade.tilab.com
26. Agent Academy, https://sourceforge.net/projects/agentacademy
27. Hall, M., Holmes, G., Frank, E.: Generating Rule Sets from Model Trees. In: Foo, N.Y. (ed.) AI 1999. LNCS (LNAI), vol. 1747, pp. 1–12. Springer, Heidelberg (1999)
28. Bishop, C.M.: Neural Networks for pattern recognition. Oxford University Press, New York (1995)
29. Aha, D.W., Kibler, D., Albert, M.: Instance-Based Learning Algorithms. Machine Learning 6, 37–66 (1991)
30. Cleary, J., Trigg, L.: K*: An Instance-Based Learner Using an Entropic Distance Measure. In: 12th Inter. Confer. on Machine learning, pp. 108–114 (1995)

A Sequence Mining Method to Predict the Bidding Strategy of Trading Agents

Vivia Nikolaidou and Pericles A. Mitkas

Aristotle University of Thessaloniki,
Department of Electrical and Computer Engineering,
Thessaloniki, Greece
+30 2310 996349, +30 2310 996390
vivia@ee.auth.gr, mitkas@eng.auth.gr

Abstract. In this work, we describe the process used in order to predict the bidding strategy of trading agents. This was done in the context of the Reverse TAC, or CAT, game of the Trading Agent Competition. In this game, a set of trading agents, buyers or sellers, are provided by the server and they trade their goods in one of the markets operated by the competing agents. Better knowledge of the strategy of the trading agents will allow a market maker to adapt its incentives and attract more agents to its own market. Our prediction was based on the time series of the traders' past bids, taking into account the variation of each bid compared to its history. The results proved to be of satisfactory accuracy, both in the game's context and when compared to other existing approaches.

1   Introduction

On-line auctions, especially the ones operated by autonomous agents, are a relatively new area of research, while still being extensively used in the real world. The subject is both wide and complex enough to guarantee an ample research potential, which is combined with the possibility to immediately deploy the obtained results to real applications. On the other hand, the information explosion during the last decade has yielded vast amounts of data, out of which useful results have to be extracted. Combined with the ever increasing processing power available, it has given a boost to the research of Data Mining techniques.

One such technique is Sequence Mining, which has generated considerable interest mainly due to its potential for knowledge extraction in large sets of sequential data. Sequence mining tries to identify frequent patterns in datasets where data items appear in a somewhat predictable fashion, such as text, aminoacid chains, or a sequence of process calls in a computer. Traditional data mining techniques, such as classification and clustering, cannot be directly applied but can be modified for sequence mining. Alternatively, sequential data can be transformed to a form that is suitable for processing by one of the existing algorithms.

The wide adoption of intelligent agents in research and practice is showing more and more examples where they are called to extract knowledge out of data, sequential or not. Agent developers can also exploit large data repositories by extracting knowledge

L. Cao et al. (Eds.): ADMI 2009, LNCS 5680, pp. 139–151, 2009.

models that can be embedded into intelligent agents [11]. Electronic auction environments represent the biggest application domain of software agents. It is in this domain where the successful blending of Intelligent Agents and Data Mining proves fruitful for both of these technologies.

In this work, we try to predict the bidding strategy of a large set of trading agents in the CAT game of the Trading Agent Competition [4]. Since the prediction is mostly based on analyzing the bids' history, we employ a novel Sequence Mining technique, which watches the variation of the bids according to their history, in order to classify the bid sequences. The architecture of the agent, for which we designed our technique, is also presented in this context.

The rest of this paper is organized as follows: Section 2 analyzes the bidding strategies that we were called to distinguish, while Section 3 presents the CAT game that was used as a framework. Section 4 explains the architecture of our agent, Mertacor, which incorporated the resulting model. Section 5 focuses on the challenging issues of Sequence Mining and subsequently analyzes our approach to this issue, eventually arriving to the results' presentation. Section 6 reviews existing literature approaches to similar issues. The work concludes with Section 7, which discusses the results and gives clues for future works.

2   Bidding Strategies

Intelligent agents, due to their autonomy and their reasoning capabilities, have long been used in auctions as bidding entities with great success. They can easily be designed to implement various bidding strategies and to adapt them, if needed. Agents can also make timely decisions that require complex computations. Although several more bidding strategies exist, the following ones are adopted in the CAT game.

2.1   Zero-Intelligence Constrained (ZI-C)

One of the most well-known strategies is the Zero-Intelligence trading strategy, first developed by Gode and Sunder [6]. This strategy randomly selects a bid based on a uniform distribution, without taking into account any market conditions or seeking any profit, hence the term Zero-Intelligence. In order to avoid possible loss, they define the Zero-Intelligence Constrained (ZI-C) strategy, used in the CAT game. This strategy sets the item's cost value as a minimum boundary to the bidding price, whereas the maximum value remains the same as in ZI-U.

2.2   Zero-Intelligence Plus (ZIP)

Since the ZI-C strategy, described in 2.1, draws bids randomly, the efficiency achieved is not enough to reach the efficiency of markets with human trading agents. As a result, Cliff in [2] introduced the ZIP strategy, in which the trading agents increase or reduce their profit margin by watching the market's conditions. More specifically, they monitor the following values: all bid and ask prices in the market, irrespective of whether the corresponding shouts led to a transaction or not, as well as the transaction price itself. The agents then adjust their profit margin accordingly.

At the beginning of a trading day, all ZIP agents have an arbitrarily low profit margin. When a transaction occurs that indicates that they could acquire a unit at a more convenient price, their profit margin is increased. However, ZIP agents have a back-up strategy, which prevents them from raising their profit margin too high. When a buyer's shout gets rejected, the shout price is increased and, when a seller's shout gets rejected, the shout price is decreased. Similarly, they watch transactions made by competing sellers and lower their profit margin if needed, so as to not be undercut by competing sellers or buyers.

2.3   Gjerstad-Dickhaut (GD)

Gjerstad and Dickhaut in [5] defined a bidding strategy based on belief functions, which indicate how likely it is that a particular shout will be accepted. This is achieved by watching the history of observed market data - namely, the frequencies of submitted bids and asks, as well as the frequencies of bids and asks that lead to a transaction. Since more recent bids and asks are of higher importance than older ones, the authors introduce a sliding window function in the agents' memory, that only takes into account the latest L shouts in the market's history.

Their belief functions are based on the assumptions that, if an ask is accepted, all asks at a lower price will also be accepted and, if an ask has been rejected, all asks at a higher price will also be rejected. Similarly, if a bid is accepted, all bids at a higher price will also be accepted and, if a bid is rejected, all bids at a lower price will also be rejected.

2.4   Roth-Erev (RE)

Roth and Erev's purpose was to create a strategy that would mimic the behavior of human players in games with mixed strategy equilibria [3]. Therefore, they used reinforcement learning algorithms on the agent's profit margins, in order to adjust them to the market's conditions.

This strategy only depends on the agent's direct feedback with the market mechanism and is therefore independent of the auction mechanism itself. More specifically, both ZIP and GD require the trading agents to have access to the history of bids and asks, as well as all accepted transactions and their prices. However, RE does not require any such data, but only relies on the same agent's interaction with the market mechanism. Therefore, it is generic enough to be used in any auction environments.

3   Reverse TAC ("CAT") Game

The CAT game is being held in the context of the Trading Agent Competition or TAC [4]. Since the markets in this game are not fixed, but instead created by each competing agent, it is called Reverse TAC, or CAT. The name CAT also refers to Catallactics, the science of economic exchange.

In the CAT game, a set of trading agents is generated by the game itself, while the contestants' purpose is to design specialist agents. Each specialist agent will operate a single market and set the rules for it. Trading agents are free to choose only one market in each operating day, and they can only buy and sell in the market that they choose.

All trading agents are either buyers or sellers and remain so for the entire game. They all buy or sell the same item in single-unit auctions. However, they are allowed to trade several items per day, placing a new bid or ask after a completed transaction.

The trading agents, buyers or sellers, incorporate one of the four trading strategies described in Section 2. Their private values, or estimations of the value of the goods traded, are drawn from a random distribution. They also incorporate a market selection strategy, which helps them choose the market that they judge to be most profitable to them. The traders' private values, bidding strategies, market selection strategies, as well as trade entitlement (the number of items that they are allowed to trade each day) are unknown to the specialists.

The specialists' goal is to design the rules and conditions of their markets. More specifically, they have to define:

a) The agent's charging policy. Agents announce their fees at the beginning of each day. They may include fees for one or more of the following:
   ⋆ Registration fee: paid by each trading agent who registers on this particular market on this day
   ⋆ Information fee: paid by each trading agent and each other specialist who requests information on the market's shouts and transaction
   ⋆ Shout fee: paid for each shout placed
   ⋆ Transaction fee: a standard sum paid for each transaction by each agent involved
   ⋆ Profit fee: a percentage on the transaction's price, paid by both the buyer and the seller
b) The market's accepting policy. This policy defines which shouts, bids or asks made by the traders are accepted and which are rejected.
c) The market's closing condition. Although a trading day consists of a number of trading rounds, announced at the beginning of the game, the market does not have to close at the end of each round, but instead is free to close at any time during the trading day.
d) The matching policy for the shouts – more particularly, which bid is matched to which ask, in order to form a successful transaction.
e) The pricing policy for each transaction. Each transaction closes at a different price, same for both the buyer and the seller.

The game consists of an unknown number of days, given in the server's configuration file but not announced. Each day consists of a number of trading rounds, with each round having a certain duration. At the beginning and at the end of each day, as well as between each day, there is some free time in order for agents to complete any possible calculations. All these durations, apart from the game's duration, are announced at the beginning of the game.

Scoring is only made during assessment days, which start and end at some random point in the game's duration. The randomness of these days, as well as the game's duration being unknown, have as a purpose to avoid exploiting initial and final conditions by the specialists, but instead focus on constructing stable markets that are able to function properly at any duration.

Scoring on each trading day and for each specialist is determined by the sum of the following three factors:

i. The agent's profit for each day, divided by the total profit of all agents for the day, in order to be normalized on a scale from 0 to 1.
ii. The agent's market share, as in, the ratio of traders subscribed to this particular market on this day, again normalized in a scale from 0 to 1.
iii. The agent's transaction rate, as in, the ratio of shouts accepted that led to a successful transaction. This is again a number from 0 to 1. Rejected shouts are not calculated.

The interested reader can find more information on the CAT game in [4].

4   Agent Mertacor

Agent Mertacor employs a modular architecture, as shown in Figure 1. Using a combination of microeconomics theory and heuristic strategies, it tries to maximize its profit without losing a significant portion of market share. The main parts of the agent are briefly discussed below.

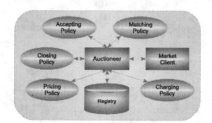

Fig. 1. Mertacor's architecture

4.1   Auctioneer

The Auctioneer is the central part in Mertacor's architecture. It is in charge of coordinating the other parts of the agent, especially when it comes to communication between the Market Client, the Registry and the policy modules of the agent. It is also in charge of computing the global equilibrium point, based on bid history. The equilibrium point is the price where the curve of offer distribution, sorted from lowest to highest, meets the curve of demand distribution. The term "global" refers to the fact that all agents are taken into account for this computation, whether they belong to the agent's particular market or not.

4.2   Market Client

The Market Client is the element that is responsible for the communication with the server. It transforms all information from the internal representation structure of the agent to the form that is understandable by the server and back. It is also responsible for subscribing to the markets of competing trading agents, in order to receive information on shouts and transactions made in their market. The reason behind this is that, normally, a specialist only has information on other specialists' profit and market share at the end of each day. Further information can only be acquired by subscribing to the other specialist's market, paying the appropriate information fee, if set.

4.3   Accepting Policy

The accepting policy is a tricky part in the agents' design. It must be low enough to keep a high transaction rate, but still high enough to guarantee many transactions. Mertacor's accepting policy is divided into two parts.

The first part is activated at the beginning of the game, before the global equilibrium point is calculated. At this point, Mertacor implements the same rule used by NYSE: namely, in order for a shout to get accepted, it must be at a price better than the day's best shout, or the day's "quote".

As soon as the global equilibrium point is calculated, Mertacor switches to an equilibrium-beating accepting policy. This means that, in order for a shout to get accepted, it must be better than the global equilibrium computed. This works as an attraction to intra-marginal traders (buyers and sellers whose private values are, respectively, higher and lower than the global equilibrium), while keeping extra-marginal traders (buyers and sellers whose private values are, respectively, lower and higher than the global equilibrium) from trading goods.

4.4   Matching Policy

Mertacor uses the matching policy described by Wurman et al in [14]. Incoming unmatched bids and asks are sorted to two separate heaps. When a bid is matched with a shout, they are automatically moved to two separate "matched" heaps. When the market is closed, the highest bids are matched with the lowest asks. This method maximizes both social welfare and Mertacor's profit margin.

4.5   Clearing Condition

Mertacor's condition for clearing the market is again twofold. In the first phase of the game, before the global equilibrium point is computed, Mertacor behaves like NYSE, clearing the market at a given probability after each shout. This method, called "continuous clearing", has been proven to be adequately efficient, while keeping a high transaction throughput [6].

After the global equilibrium point calculation is completed, Mertacor switches to a modified round-clearing condition. This means that the market is closed at the end of each round. The variation used is that Mertacor switches again to a continuous clearing policy towards the end of the game, in order to maximize throughput.

4.6   Pricing Policy

Like the accepting policy and the clearing condition, Mertacor's pricing policy is also divided into two phases: before and after the global equilibrium point is computed.

During the first phase, Mertacor uses a variation of a discriminative k-pricing policy [10]. The k parameter is computed as the ratio of sellers in this game, resulting in a k of 0.5 for a balanced market. Our policy slightly favors sellers when there are more buyers and favors buyers when there are more sellers, giving them an incentive to balance the market.

In the second phase of the game, Mertacor switches to a global equilibrium pricing policy. This means that the price for all transactions is the same price as the global equilibrium. This way, Mertacor gives each trader the same profit that they would gain in an efficient global partitioning.

In both cases, though, Mertacor favors intra-marginal traders at the expense of extra-marginal ones. This means that, if a transaction is executed between an intra-marginal trader and an extra-marginal one, Mertacor will clear it at the price given by the intra-marginal trader.

4.7   Charging Policy

Choosing the right charging policy is the most challenging task in the design of a CAT agent. Setting the fees charged too low may not yield enough revenue, while values that are too high may discourage traders from joining the specialist's market.

As explained in Section 3, each specialist may impose five different fees: a) registration fee, b) information fee, c) shout fee, d) transaction fee, and e) profit fee.

Mertacor only imposes the profit fee, keeping the other ones down to zero. This is decided keeping in mind that only successful transactions must lead to a payment – in other words, an agent who does not earn anything should not pay anything either.

Heuristic experiments showed that the optimal value (i.e., the value which maximizes the specialist's score) is 0.2. This can go as low as 0.1, whenever the specialist's market share is too low, but as high as 0.3, when the specialist estimates that the market conditions are suitable.

5   Predicting the Bidding Strategy

In order to classify each trading agent as intra-marginal or extra-marginal, we needed to know its private value. The most crucial step in determining it was to find the agent's bidding strategy, which had to be predicted by observing its bidding history.

5.1   Modeling the Problem

The problem of predicting the bidding strategy of an agent can be seen as a classification problem, where each agent has to be assigned one of four labels (ZI-C, ZIP, GD or RE). However, it differs from conventional classification problems, in that the most characteristic input data is the past history of bids. With each bid having a specific timestamp and being correlated to its past bids, the problem is identified as a sequence mining one, where various time series have to be matched into labels.

Sequence mining is of particular interest because of various reasons. The first is the high dimensionality of the problem. More particularly, in our case, the history of past bids can comprise several hundreds or even thousands of samples, depending on the game's duration. One can easily understand that it is necessary to reduce the problem's dimensionality, allowing for better performance of the classification algorithms.

However, high dimensionality is but the tip of the iceberg. In many cases, including ours, the number of dimensions is not known a priori, but instead continuously grows according to the bids history. For example, during the first trading days of the game, we often have fewer than ten past samples. However, as the game advances and bids are constantly added to the history logs, the time series grows longer.

The third and most important issue in sequence mining is the notion of "sequence". This means that it is not merely a set of incoming value-timestamp pairs, but each value and each timestamp is related to its history. In fact, most information can be deduced by looking at the bid history and not at a bid itself, since, for example, ZI-C could give practically any bid value observed separately. On the other hand, conventional classification or clustering algorithms treat each feature as independent of the other ones and do not look for relationships between features. This means that it is not easy to examine each bid's history.

The most common approach in bibliography so far, for example by Martinez-Alvarez et al [9], is to use a moving window to split the time series and feed the latest N samples to the classifier. The authors take this approach one step further by normalizing input data. However, this still does not provide a satisfactory solution to the third issue described above, since each sample is still treated independently in the algorithms.

5.2 Our Approach

In our approach, we take this philosophy one step further by introducing deltas to the classifier. More specifically, let us consider a set of prices p_0... p_n, with p_0 representing the most recent one, and their corresponding set of timestamps, t_0... t_n. Notice that increasing subscript values in the sequence denote earlier points in the timeline. We define a sliding window of size 6, but only the first value p_0 is fed as-is to the classifier's input. For each of the rest of the values p_1 to p_5, we subtract the previous one in the series (more recent in the bid history). The results, d_0 to d_4, give the classifier the notion of sequence, therefore facing the corresponding issue.

Furthermore, we have four additional features as the classifier's inputs. These are calculated by subtracting, for each one of d_1 to d_5, the value of the second previous sample, d_0 to d_4 accordingly.

Since the difference in time is more independent statistically than the difference in bid value, we only define a simple sliding window of size 3 when it comes to time differences. This means that we have three time-related attributes: t_0, $t_0 - t_1$, and $t_1 - t_2$.

The input dataset is completed by adding two more attributes independent from the time series, namely the type of the trader (buyer or seller), as well as whether this particular bid led to a successful transaction.

Since agents have limited time available for calculations while the game is running, the model was built offline. Classification was continuous, with the traders being assigned a label multiple times during the course of the game. In the case of a wrong prediction, traders' private value would be estimated wrongly, which might eventually lead to their misclassification as intra-marginal or extra-marginal.

Training data was initially taken from our own experiments with the CAT platform. This was used in the first version, designed for the qualifying rounds of the game. However, for the final games, we retrained the model using values from the qualifying games only. Only qualifying games 2 and 4 were used. The reason behind this is that games 1 and 3 were not run at full-length, but instead lasted only 100 trading days, which is only barely enough for the traders to finish exploring the markets and settle to their preferred one. While the number of trading days is unknown due to the game specifications, it was hinted that they would last much longer. In any case, since our model's inputs only depended on the data so far, the algorithm was designed to perform well in full-length as well as slightly shorter games.

Algorithm Selection. The next step was the selection of the most suitable algorithm. This is done in a manner similar to [12], but skipping the data preprocessing part. We compared the following three algorithms: Neural Networks, J48 and Support Vector Machines. We used the WEKA platform [13] to train the model, using 10-fold cross validation.

We eventually decided to discard Support Vector Machines because of its slow response and inadequate performance for the specific input data. Time was an important issue, since the trading rounds are of fixed duration. Additionally, the classification accuracy did not exceed 30%. Meta-classification might have slightly improved it, but it might have been at the expense of complexity.

Of the remaining two algorithms, J48 outperformed Neural Networks, giving a classification accuracy of 56.5 percent. Confirming our findings in [12], meta classification, in the form of Bagging, further enhanced the model's performance, boosting it to 65.5 percent. Final results are presented in Table 1.

Table 1. Summarized classification results

Algorithm	Correctly Classified	Incorrectly Classified	Kappa Siatistic	Mean abs. error	RMS error
Bagging - J48	65.55%	34.45%	0.54	0.24	0.34
J48	56.47%	43.53%	0.42	0.23	0.44
SMO	29.22%	70.78%	0.03	0.36	0.46
Perceptron	45.45%	54.55%	0.26	0.33	0.41

Tables 2 amd 3 presents the detailed accuracy by class, as well as the confusion matrix, for the winning algorithm, the combination of Bagging and J48.

The Effect of Difference Order. In our dataset, we used differences of first and second order, which are a rough representation of the first and second derivative. In order to further illustrate the effect of difference order in the results, we constructed three datasets: price1, price2 and price3, which contain the differences of first, first and second, and first to third order, respectively. The chosen algorithms were run on all three datasets and the comparative results are depicted in Table 4.

Table 2. Detailed accuracy by class for winning algorithm

TP Rate	FP Rate	Precision	Recall	F-Measure	ROC Area	Class
0.65	10.3%	68.0%	0.66	0.67	0.87	GD
0.72	13.7%	65.7%	0.73	0.69	0.89	ZIC
0.61	16.2%	58.8%	0.61	0.60	0.83	ZIP
0.61	6.2%	72.8%	0.62	0.67	0.88	RE

Table 3. Confusion matrix for the winning algorithm

GD	ZIC	ZIP	RE	< − − classified as
4936	536	1296	739	GD
326	**5787**	1533	107	ZIC
1210	1562	**5034**	417	ZIP
782	926	700	**3909**	RE

Table 4. Classification accuracy by algorithm and difference order

	price1	price2	price3
Perceptron	43.37	45.45	45.00
SMO	29.26	29.22	29.25
J48	56.66	56.47	55.76
Bagging − J48	65.21	65.55	65.67
Average	48.63	49.17	48.92

We observe that the dataset which contains the second-order difference has the highest average classification accuracy on this dataset. For the highest-performing algorithm, namely the combination of Bagging and J48, we observe a slightly better performance for the price3 dataset. However, since all differences are computed on the fly during the course of the game in order to obtain a correct classification, we decided to use the price2 dataset, which also gives the highest average performance and needs less computational resources than price3.

6   Related Work

The most relevant work in bibliography is the one of Gruman and Narayana in [7]. They compared the performance of Support Vector Machines (SVMs) and Hidden Markov Models (HMMs). Their experiments were made in benchmark-like tests, using traders and specialists of controlled variations. They obtained a classification accuracy ranging from 52 to 62 percent using HMMs, according to fine-tuning of the HMM's parameters. Our approach not only outperforms it, but it is also implemented in "real-game" instead of controlled conditions. Furthermore, in contrast to HMMs, the decision-tree-based J48 hardly requires any fine-tuning, making it more robust when tested in the real world.

Bapna et al in [1] attempt to predict trading agents' Willingness-To-Pay in auction environments. When it comes to classifying agents' bidding strategies, they assume a single bidding strategy for all traders, namely the MBR (Myopic Best Response) strategy, and classify traders as Evaluators, Participators and Opportunists, as well as MBR and non-MBR traders. Details on the classification method, the input data, as well as classification accuracy, are not given, since the work's focus is mainly on Willingness-To-Pay and trader classification is merely used as an intermediate step.

When it comes to sequence classification, one example is the work by Lesh et al in [8]. They present an algorithm which examines all features of the given sequence and selects the ones to be used as input to traditional classification algorithms, such as Na?ve Bayes or Winnow. Zaki enhances this work in [15] by taking into account issues, such as length or width limitations, gap constraints and window size. This approach has increased complexity, but still does not take into account the continuity of the time series.

The sliding window approach prevails in more recent works, such as Martinez-Alvarez et al in [9]. They have a sequence clustering problem, in which they normalize the input data and subsequently determine the optimal window size. Taking into account the similarities between classification and clustering, this approach further supports the novelty of our work taking into account deltas instead of a sliding window, even normalized.

7 Conclusions and Future Work

We have used sequence mining, a data mining technique, to improve the decision mechanism of an agent. Using this technique, a specialist agent can successfully predict the bidding strategy of a set of trading agents. Our model was designed as part of a widely accepted, general-purpose trading agent competition game. Nevertheless, it remains generic enough to allow its adaptation to a multitude of generic auction environments with little or no modification.

The classification accuracy proved to be satisfactory, since our agent was able to correctly classify a large majority of the traders. Using J48, an algorithm based on decision trees, also had the advantage of quick performance, allowing Mertacor more time for other calculations.

Mertacor proved competitive by ranking fifth in the final games out of a total of 14 participants. Games were often augmented by the eventual presence of one or more test agents.

Future work in this direction will most likely be headed towards the direction of further introducing the notion of continuity into the models. The main current problem is that, while there is only one correct class for each time series, our model makes several predictions, one for each bid placed. Results could probably be enhanced by a meta-layer, which would take into account previous predictions of the same model for each agent and decide accordingly.

Acknowledgements

This work is part of the 03ED735 research project, implemented within the framework of the Reinforcement Programme of Human Research Manpower (PENED) and cofinanced by National and Community Funds (25% from the Greek Ministry of Development-General Secretariat of Research and Technology and 75% from E.E.-European Social Funding).

References

[1] Bapna, R., Goes, P., Gupta, A., Karuga, G.: Predicting Bidders' Willingness to Pay in Online Multiunit Ascending Auctions: Analytical and Empirical Insights. Informs Journal on Computing 20(3), 345–355, INFORMs (2008)

[2] Cliff, D.: Minimal-intelligence agents for bargaining behaviours in market-based environments. Technical Report HP-97-91. Hewlett-Packard Research Laboratories, Bristol, England (1997)

[3] Erev, I., Roth, A.E.: Predicting how people play games: Reinforcement learning in experimental games with unique, mixed strategy equilibria. American Economic Review 88(4), 848–881 (1998)

[4] Gerding, E., McBurney, P., Niu, J., Parsons, S., Phelps, S.: Overview of CAT: A market design competition. Technical Report ULCS-07-006. Department of Computer Science, University of Liverpool, Liverpool, UK, Version 1.1 (2007)

[5] Gjerstad, S., Dickhaut, J.: Price formation in double auctions. Games and Economic Behaviour 22, 1–29 (1998)

[6] Gode, D.K., Sunder, S.: Allocative efficiency of markets with zero-intelligence traders: Market as a partial sustitute for individual rationality. Journal of Political Economy 101(1), 119–137 (1993)

[7] Gruman, M.L., Narayana, M.: Applications of classifying bidding strategies for the CAT Tournament. In: Proceedings of the International Trading Agent Design and Analysis Workshop, TADA 2008, Chicago, IL, USA, July 14, pp. 11–18. AAAI, Menlo Park (2008)

[8] Lesh, N., Zaki, M.J., Ogihara, M.: Mining features for sequence classification. In: Proceedings of the fifth ACM SIGKDD international conference on Knowledge Discovery and Data Mining, San Diego, CA, United States, August 15-18, pp. 342–346. ACM, New York (1999)

[9] Martinez – Alvarez, F., Troncoso, A., Riquelme, J.C., Aguilar – Ruiz, J.S.: LBF: A Labeled-Based Forecasting Algorithm and Its Application to Electricity Price Time Series. In: Proceedings of the Eighth IEEE International Conference on Data Mining, ICDM 2008, Pisa, Italy, December 15-19, pp. 453–461. IEEE, Los Alamitos (2008)

[10] Sattherthwaite, M.A., Williams, S.R.: The Bayesian Theory of the k-Double Auction. In: Friedman, D., Rust, J. (eds.) The Double Auction Market – Institutions, Theories, and Evidence, ch. 4, pp. 99–123. Addison-Wesley, Reading (1993)

[11] Symeonidis, A.L., Mitkas, P.A.: Agent Intelligence through Data Mining. Springer Science and Business Media (2005)

[12] Symeonidis, A.L., Nikolaidou, V., Mitkas, P.A.: Sketching a methodology for efficient Supply Chain Management agents enhanced through Data Mining. International Journal of Intelligent Information and Database Systems 2(1), 49–68 (2008)

[13] Witten, I.H., Eibe, F.: Data Mining: Practical Machine Learning Tools and Techniques with Java Implementations. Morgan Kaufmann, New Zealand (1999)

[14] Wurman, P.R., Walsh, W.E., Wellman, M.P.: Flexible double auctions for electronic commerce: theory and implementation. Decision Support Systems 24(1), 17–27 (1998)

[15] Zaki, M.J.: Sequence mining in categorical domains: incorporating constraints. In: Proceedings of the ninth international conference on Information and knowledge management, McLean, Virginia, United States, November 6-11, pp. 422–439. ACM, New York (2000)

Part IV

Agent Mining Applications

Agent Assignment for Process Management: Pattern Based Agent Performance Evaluation

Stefan Jablonski and Ramzan Talib

University of Bayreuth, Applied Informatics IV, D-95440 Bayreuth, Germany

Abstract. In almost all workflow management system the role concept is determined once at the introduction of workflow application and is not reevaluated to observe how successfully certain processes are performed by the authorized agents. This paper describes an approach which evaluates how agents are working successfully and feed this information back for future agent assignment to achieve maximum business benefit for the enterprise. The approach is called Pattern based Agent Performance Evaluation (PAPE) and is based on machine learning technique combined with post processing technique. We report on the result of our experiments and discuss issues and improvement of our approach.

Keywords: Agent Assignment, Workflow, Process, Machine Learning, Business Benefit, Pattern base Agent Performance Evaluation.

1 Introduction

Many applications can be described by processes. For example, in a garment factory a garment production process starts with the assignment of designers to design tasks (e.g. to design a new T-shirt), continues with the design of a new garment, continues with its production, and finally ends up with selling the new products. Of course, from a business (benefit) perspective there is a close relationship between the multiple steps of such a process. We focus in this article the dependency of good sales figures from the assignment of designers to design the tasks. Substantiated by our experience with garment production it is obvious that certain designers are delivering excellent designs for specific goods, while other designers only deliver moderate designs for these goods which lead to only moderate sales figures. The main two questions in the context of this observation are:

- How to get aware of dependencies like the above one?
- How to respond to such an observation?

We want to discuss these two questions in different applications contexts: the first one is the conventional one, where designer assignment is done by responsible managers; the second one is characterized by the deployment of workflow management systems which automatically assign designers to design task.

In the conventional case it depends on the experience and the analytical knowledge of responsible managers to assign the right staff to the right tasks: due to intensive observations managers find out which staff (agent) assignment results in best business

L. Cao et al. (Eds.): ADMI 2009, LNCS 5680, pp. 155–169, 2009.

benefit . When managers recognize that specific persons are guaranteeing best business benefit they consider this in future assignments. However, often these observations are not substantiated by analyzing production and sales data but are heavily determined by personal experiences and observations of responsible managers.

When automatic staff assignment like in workflow management systems takes place, personal experience is normally not incorporated into staff assignment. People are assigned to process steps if they are eligible to perform such a step. Whether these eligible persons are producing good or moderate sales figures can normally not be considered at the assignment [9]. In Section 2.2 we will detail staff assignment in workflow management; this will back up our observation.

In order to improve staff assignment two things have to be accomplished. First, business intelligence has to be introduced to detect dependencies between staff assignment and business benefit. In the conventional case, responsible managers could exploit this knowledge to perform more adequate staff assignment. In the case of automatic task assignment the results of business intelligence must be fed back into automatic task assignment policies.

In this paper we want to focus automatic task assignment of workflow management systems since it requires a kind of superset of solution strategies as manual task assignment needs. Concretely, we firstly want to identify effective and efficient forms of business intelligence for staff assignment in workflow management applications. Secondly, we want to define a method how the findings of business intelligence can be incorporated into staff assignment policies to produce more productive staff assignments which leads to better business performance.

Next, a short example will be depicted which will illustrate the dependency between agent assignment and the success of process execution. Again, we want to refer to a scenario from garment industry. Agent assignment is about the assignment of designers to special design tasks. The success of performance of a business process as a whole and the designer assignment in detail is measured by sales figures. Hereby, it has to be mentioned that business benefit here is in the form of good sales figures.

How does agent assignment work? Before agents can be assigned to processes (in order to perform them) they are characterized and classified. For instance, some employees of a garment company are classified as designers; other employees are classified as sale assistants or purchasers. For each step of a process eligible agents are determined. For example, designers are assigned to the process step *Create Design*; sales assistants are assigned to the process step *Define Selling Strategy*.

Consider a process *Create Design* of a garment enterprise process. Agents who selected this process step are responsible to create a specific design (e.g. shirt, pants). Furthermore, designs might be classified into different types. For example, shirt might be specialized into the sub-classes T-shirt and Blouse. The type of design that is needed is determined by some input data for the *Create Design* process step such that the respective designers knows what kind of design has to be created.

Each design created by an agent in the *Create Design* process step has different sales values (e.g. GOOD, MODERATE). Good sales value of a design depends on the quality of design and therefore directly on the performance and expertise of the assigned designer, e.g. the agent assigned to the process *Create Design*.

Table 1. Data pattern and agent performance

| | Shirt Patterns | | | |
| | T-shirt | | Blouse | |
Business benefit	GOOD	MODERATE	GOOD	MODERATE
Gabi	84	6	13	72
Clara	19	78	38	5

We now want to correlate agent assignment and products that had to be designed and do achieve business benefits. Table 1 depicts this correlation. Statistics in Table 1 shows that Gabi has good expertise of creating T-shirt designs (84 designs for T-shirt out of 90 designs result in good business benefits, i.e. good sales values) while Clara has expertise of creating Blouse designs. Besides, Gabi's expertise in designing Blouse is rather moderate (most of his designs result in moderate sales values); almost the same holds for Clara's design for T-shirt. Relating the results of Table 1 to our two contributions the following two tasks can be identified:

- It is necessary to detect relationships between type of designs (design patterns), agent assignment and achieved business benefits . We apply machine learning techniques to accomplish this task.
- It is necessary to provide a feedback cycle between encountered design patterns and future agent assignments.

Design patterns are general design groups which predict business benefit more accurately. They can be expressed as an expression of attributes-values of design data. Machine learning technique can be used to learn data patterns from any dataset. For the example from Table 1 this means: Identify relationships between different design patterns, agent assignments and achieved business benefits, first. Then take this knowledge and provide it for further selections of agents for specific design tasks.

Currently in WFMS, agents assignments are usually defines once and are not reevaluated whether agents do their job good or bad. From a business point of view our approach leads to two major benefits: First, relationships between data patterns and agent performance can be detected. This result is valuable for both conventional agent assignment and for agent assignments performed by WFMS. Second, in the latter case our approach guarantees automatic agent assignments promising best business benefit. Again, it is completely application dependent how business benefit is defined. This has to be done by domain experts before our approach is applied.

We call our approach Pattern based Agent Performance Evaluation (PAPE). It consists of five major phases: In the first phase, domain knowledge is incorporated to find how business benefits are defined in the application. Second phase generates integrated data structure combining the machine learning technique and post-processing technique. This integrated data structure is used to evaluate the dependency between agent assignment and success of process executions in the third phase. If a dependency is observed agent assignment rules are learned in the fourth phase and are presented to domain expert for further refinement. Rules finally refined by domain expert are updated in the WFMS (fifth phase).

As a prerequisite we need a sort of an event log (audit trail) which stores both data about agent assignments and achieved sales figures. Most information systems like WFMS, ERP, CRM, and B2B systems provide such event logs [13]. Especially, we will concentrate in this paper on the application of our method to WFMS which typically maintain such a log.

The reminder of the paper is organized as follows. We provide a general overview on workflow management systems and staff assignment strategies in Section 2. Section 3 explains our methodology. In Section 4 we are presenting the result of applying our method on real data sets. We discuss related work in Section 5. Section 6 concludes this paper and provides an outlook on future work.

2   Workflow Management

2.1 . Overview

Workflow management is a technology which aims at the execution of workflow steps. It is concerned with providing the right information to support each step. Also programs and tools, needed to perform the work steps, are provided by agents who are determined by a workflow management system as well. Finally data are routed between work steps that are either consumed and/or produced by them.

In this article we use the terms "process" and "workflow" synonymously though we principally see a difference between them [5]. However, in this article we always concentrate on the execution aspect and therefore want to take them as synonyms.

According to [5] a process consists of five major perspectives which altogether define how processes are executed. The functional perspective defines the skeleton of a process: it identifies the process and comprises general information about the process. Also, the functional perspective determines whether the process is elementary or composite. The behavioral perspective determines the control flow, i.e. the order of process execution. The operational perspective describes tools (programs, systems etc.) that are available for the execution of a process. The data perspective defines input data and output data of a process. This perspective consequently then determines data flow within a process. Last but not least the organizational perspective determines agents who are eligible to perform a certain process. This task is usually called agent assignment [2].

For this article the organizational perspective is the most interesting one. Nevertheless, our research could also be applied to other perspectives of a workflow. We already have investigated this issue; it will be presented in future articles.

The main task of a workflow management system is to coordinate all mentioned perspectives when a process has to be performed. In Fig. 1 a small example is depicted.

Fig. 1. Create design process in garment factory

It shows two out of many processes that are executed during the design process. In the first step shown the garment collection (e.g. for the next season) has to be determined. The agent "Marketing" (i.e. the corresponding department) is responsible for that. It uses an ERP application in order to define the next collection. The output of this process is "Design Specification" which is a compound data item that describes all designs that have to be done for the next garment collection. This data is passed as input to the next step "Create Design". Here designers (agents) have to create designs utilizing a so called application "DesignApp".

A process management system has to take care that all process steps are executed in the right order, consuming the right input data, producing the right output data, applying the right applications, and last but not least select the right agents who have to perform a step. This last task will be considered in detail in the next sub-section.

2.2   Overview of Agent Assignment Strategies

In order to determine agents who have to perform a certain process the following steps are executed. First, agents are classified according their capabilities. Normally, they are associated to certain roles (e.g. designer, clerk). Second, those roles are used to describe a process step (organizational perspective). When a process step is executed its organizational perspective, i.e. the associated role is taken and all agents that are corresponding to such a role are informed about work to do. One of the eligible agents (mostly the fastest one) then claims that piece of work and has to perform it.

Here we present a short example of agent assignment rules for the **Create Design** process we are using throughout this article. The first rule shows which agents belong to the **Designer** role; the second rule shows that agents with the Designer role can execute the **Create Design** process.

- Role Designer [] = { "Gabi", "Clara", "Jan", "Eva", "Bali", "Emil" } ;
- Performer.Role() = Designer;

In almost all workflow management systems the role concept is coarse grained (i), is determined once at the introduction of the process application (ii), and is not integrated in a feedback loop (iii). The first argument means that roles are provided as "designer" or - and this is preferable - "Shirt Designer", "Pants Designer", etc. The second argument means that these role associations are done once and are not updated continuously. The third and most important argument means that it is not evaluated, i.e. it is not observed how successful certain tasks are performed in order to improve role assignment. For example, if a "Shirt Designer" would always produce designs which are very bad, (s)he should be deprived of this role. All in all, agent assignment in conventional process management systems could be characterized as static.

Now, it is appropriate to re-state the contribution of this paper. First, we contribute an approach to determine how agents are working successfully, i.e. whether they are contributing to the business benefit . Second, we feed this information back into upcoming agent assignments , i.e. we associate those agents with process step who (most probably) guarantee a successful outcome of the whole process.

3   PAPE System Architecture

The proposed PAPE method evaluates dependency between data patterns, agent assignment and business benefit to learn agents working successfully and feeds back this knowledge for future agent assignments to achieve maximum business benefits.

How to evaluate the dependency of agent assignment and business success? Classical data mining techniques are data centric and solve business issues in an isolated way [3]. Thus, often impact of other domain resources or resource interdependency is lost. For the PAPE method, machine learning classifier can easily be employed to determine how design patterns predict business benefit (GOOD or MODERATE) but cannot directly be employed to evaluate the consequence of agent's expertise for GOOD or MODERATE business benefit. Moreover, for the respective design pattern, these classifiers cannot predict best agent promising best business benefit.

So, to subjugate this problem and to learn dependency of agent assignment and business benefit the PAPE method uses machine learning technique (decision tree) combined with post-processing technique to generate an integrated data structure. Our method uses the workflow audit trail for a particular time period and generates integrated data structure which is used to evaluate the dependency between data patterns, agent assignments and business benefit . If this dependency is observed then performance based agent assignment rules are learned. These rules are more productive and are provided as a feedback for automatic agent assignments in WFMS which leads to successful process execution and better business performance.

The PAPE method is shown in Fig. 2. It consists of five phases: Preprocessing Phase (Section 3.1), Integrated Data Structure Generation Phase (Section 3.2), Pattern based

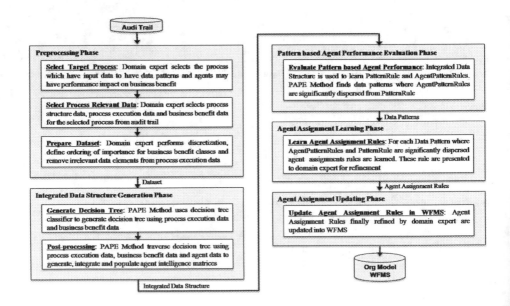

Fig. 2. PAPE method

Agent Performance Evaluation Phase (Section 3.3), Agent Assignment Learning Phase (Section 3.4) and Agent Assignment Updating Phase (Section 3.5).

3.1   Preprocessing Phase

In the preprocessing phase the workflow log is processed by domain expert and domain knowledge is incorporated to prepare the dataset to be used by PAPE method. Since it is completely application dependent how business benefits are defined, domain knowledge is required. Domain knowledge is used to decide different issues which are discussed in the following sub-sections as a guide line for preparation of dataset suitable for the PAPE method:

- Which processes should be evaluated by PAPE Method?
- Which data elements (e.g. sale) in the audit trail define a business benefit ? If data values (sale) are not nominal then which range of values belongs to different business benefit classes (e.g. GOOD, MODERATE)?
- Which business benefit class is interesting (e.g. GOOD)?

Select Target Process: In order to find out which process is suitable for the PAPE approach, the following questions have to be answered:

- Does a process have available input data to get different data patterns?
  → Only if data patterns can be identified such a process qualifies for PAPE.
- Do agents have different expertise or performance impact on business benefits for a process? → Only if different expertise profiles are predominant agent selection qualifies for PAPE.

Domain expert can answer these questions. The *Create Design* process has input data (design specification) that form data patterns and the business benefit is depending on the performance of executing agents as statistics from Table 1 reveals. Domain expert decide that *Create Design* is a target process suitable for PAPE.

Select Process Relevant Data: Data required for PAPE consist of process structure data, process execution data and business benefit data. Process structure data consist of process name, agent role and executing agent information. Process execution data consist of different data elements (e.g. design features) and is input to process before execution starts. Business benefit data is either part of execution data or is available somewhere in the audit trail and is used for the evaluation of process success as a whole and agent performance in detail.

For the selected process domain experts select the process structure data , process execution data and business benefit data. It is application dependent how business benefits are defined. Business benefit data may be the part of application data or process execution data . This business benefit data should be integrated and consistent with process execution data . Domain experts extract and integrate business benefit data with process execution data and process structure data. For instance in *Create Design* process, business benefit data (e.g. sale) is the part of marketing department application data source. The relation between a process and its business benefit cannot be inferred automatically. Therefore use of domain knowledge is necessary. Domain experts prepare this dataset consisting of process structure data, process execution data and business benefit data.

Prepare Dataset: The last step of the Preprocessing Phase is to prepare the dataset which is acceptable for next phase. This phase applies machine learning technique to the input dataset to generate an integrated data structure. Data suitable for machine learning technique needs to be in nominal or numeric form. Process structure data is already in nominal form (e.g. Agent, Role, Process Name) but some data elements of the process execution data may not fulfill this requirement so need to be transformed into nominal form or ignored (e.g. design name). In general business benefit data is already in the nominal form.

If the business benefit data is not in discrete or nominal form then domain experts transform this business benefit data into discrete classes like GOOD, MODERATE. Discrete classes are defined by the domain expert as required in application context like EXCELLENT, GOOD, AVERAGE and POOR. When the dataset is preprocessed, the last step required is to define interesting business benefit class. Domain experts define when dataset is loaded into PAPE method and used in phases: Pattern based Agent Performance Evaluation and Agent Assignment Learning.

3.2   Integrated Data Structure Generation Phase

The PAPE method generates integrated data structure consisting of a decision tree and agent intelligence matrices using the dataset prepared in the preprocessing phase. It is used to identify the relationship between data patterns, agent assignments and business benefits . To generate an integrated data structure, first a decision tree is generated from process execution data and business benefit data using machine learning technique. Then post-processed is performed to extend the decision tree with agent intelligence matrices (one matrix for each branch of the tree) using process execution data, agent data and business benefit data.

Generate Decision Tree: A decision tree is generated using the J48 classification algorithm selected from Weka library [4]. It is a slight modification of the C4.5 decision tree [11]. It considers all the possible tests that can split the dataset and select a test set that gives the best information gain. It generates a decision tree for the given dataset by recursive partitioning and can handle nominal or numeric values.

The J48 algorithm is applied on process execution data (e.g. design features) and business benefit data (as classifying attribute). The algorithm generates a decision tree that consists of branches and leaves (Fig. 3 (A)). Each branch represents a data pattern described by process execution data (e.g. MajorType="Shirt" and MinorType="Blouse"); the leaves represent the predicted business benefit class (e.g. GOOD) and accuracy of prediction (e.g. 50/12: which means out of 50 design instances 38 are GOOD and 12 are not GOOD).

Figure 3(A) displays the decision tree generated from the dataset of the *Create Design* process which is actually of large size; however, to demonstrate our method more precisely we have selected only a partial tree. Now, it typically constitutes 4 data patterns learned by 213 designs instances created by 6 agents of the *Create Design* process.

It needs to be said that J48 algorithm may generate a decision tree consists of single node (no branches) then PAPE method is not applicable for such processes because if no data patterns are found then relationship between data pattern, agent assignment and business benefit.

Post-processing: If the decision tree consists of different branches (data patterns) then post-processing is performed to extend each branch of the decision tree with an agent intelligence matrix. An agent intelligence matrix (Fig. 3 (B)) is a two dimension array whose number of rows and columns are determine by number of agents and business benefit classes in the dataset respectively and each particular row-column entry can be identified by agent and business benefit data.

To extend each branch of the decision tree with agent intelligence matrix the post-processing scenario is described: For each instance of the dataset, process execution data is used to parse the decision tree by traversing a particular branch and find its agent intelligence matrix. If the agent intelligence matrix already exists then only its row-column entry is incremented otherwise an intelligence matrix is extended with the branch and then the row-column is incremented. Which row-column entry of the agent intelligence matrix needs to be incremented? It is determined by the agent and business benefit data value of the instance being processed. When all instances of the dataset are processed an integrated data structure is generated as shown in Fig. 3.

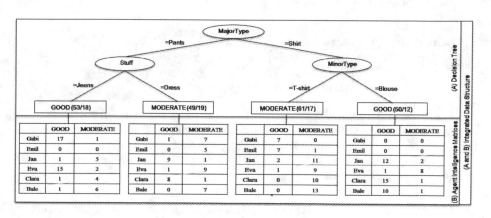

Fig. 3. Integrated data structure

3.3   Pattern Based Agent Performance Evaluation

The goal of this phase is to evaluate either all agents execute the process almost equally well (achieve business benefit) or different expertise profiles are predominant? For this evaluation there is need to first determine dependency between data pattern, agent assignment and business benefit and then perform pattern based agent performance evaluation.

Pattern based agent performance evaluation is performed using the integrated data structure. The decision tree can be used to evaluate dependencies between data pattern and business benefit (Fig. 3 (A)); the decision tree extended with agent intelligence matrices (Fig. 3 (A and B)) can be used to evaluate dependency between data pattern, agent assignment and business benefit . Hence, from the integrated data structure two types of rules can be learned: PatternRule and AgentPatternRule. PatternRule relates the data pattern with a business benefit class (from root to leaf of the tree Fig. 3 (A)); the AgentPatternRule relates data pattern, agent (assignment) with a business benefit class (from root to particular row-column entry in agent intelligence matrix Fig. 3 (A and B)). For instance consider the two rules from rightmost branch of the integrated data structure:

- (MajorType= "Shirt" and MinorType= "Blouse")
  $\rightarrow$ Sale = GOOD [Accuracy 76%]
- (MajorType= "Shirt" and MinorType= "Blouse" and agent="Clara")
  $\rightarrow$ Sale = GOOD [Accuracy 94%]

The improved accuracy of the AgentPatternRule in comparison to the PatternRule reveals that the agent (e.g. Clara) has expertise for the data pattern: there is a dependency between data pattern, agent assignment and business benefit. This agent expertise can be utilized for future agent assignment.

If on the other hand accuracy of the PatternRule rule is greater than the AgentPatternRule rule then agent has no significant expertise for this data pattern. So there is no dependency between data pattern, agent assignment and business benefit. This is shown in the following rules:

- (MajorType= "Shirt" and MinorType= "Blouse")
  $\rightarrow$ Sale = GOOD [Accuracy 76%]
- (MajorType= "Shirt" and MinorType= "Blouse" and agent="Eva")
  $\rightarrow$ Sale = GOOD [Accuracy 11%]

This agent expertise cannot be utilized for future agent assignment for achieving business benefit. It needs to be mentioned that only an interesting business benefit class (e.g. GOOD) is used for rules learning and evaluation because we are interested in utilizing agent expertise to promote business for GOOD sale. This interesting business benefit class is mentioned by domain expert when dataset is loaded into the PAPE method (Section 3.1).

This phase finds those data patterns where agents have significant expertise in their attached agent intelligence matrix. When accuracy of AgentPatternRules are within $\alpha = 0.05$ range of PatternRule then agents are working almost equally well. Here $\alpha = 0.05$ is an acceptable dispersion level and can be adjusted by domain expert as and when required. A list of data patterns is determined where AgentPatternRules are significantly dispersed from PatternRule and forwarded to next phase for learning agent assignment rules.

3.4   Agent Assignment Learning Phase

For each pattern where a relationship between data pattern, agent assignment and business benefit is observed this phase generates agent assignment rules. A list of AgentPatternRules whose accuracy is significantly ($\alpha = 5$) higher than PatternRule are learned from the respective agent intelligence matrix. For the integrated data structure in Fig. 3 the following patterns and respective agent assignment rules can be learned:

- (MajorType="Shirt" and MinorType="Blouse") [Accuracy = 76%]
  - IF (Agent="Clara") → (Sale="GOOD") [Accuracy = 94%]
  - IF (Agent="Bale") → (Sale="GOOD") [Accuracy = 91%]
  - IF (Agent="Jan") → (Sale="GOOD") [Accuracy = 86%]
- (MajorType="Shirt" and MinorType="T-shirt") [Accuracy = 28%]
  - IF (Agent="Gabi") → (Sale="GOOD") [Accuracy = 100%]
  - IF (Agent="Emil") → (Sale="GOOD") [Accuracy = 88%]
- (MajorType="Pants" and Stuff="Dress") [Accuracy = 39%]
  - IF (Agent="Jan") → (Sale="GOOD") [Accuracy = 90%]
  - IF (Agent="Clara") → (Sale="GOOD") [Accuracy = 89%]
- (MajorType="Pants" and Stuff="Jeans") [Accuracy = 66%]
  - IF (Agent="Gabi") → (Sale="GOOD") [Accuracy = 94%]
  - IF (Agent="Eva") → (Sale="GOOD") [Accuracy = 88%]

Agent assignment rules are learned and are presented to domain experts for refinement. Rules finally refined by domain expert are forwarded to the Agent Assignment Updating Phase.

3.5   Agent Assignment Updating Phase

Agent assignment rules finally refined by domain experts need to be deployed in the organizational model of the WFMS. This phase transforms assignment rules into XML form which can be deployed in a WFMS. For each data pattern and its expert agents, a role is created and agents are defined for each role. For instance consider the following notions for agent assignment rules learned in the previous sub-section:

- Role JeansPantsDesigner [] = {"Gabi", "Eva"};
- Role DressPantsDesigner [] = {"Jan", "Clara"};
- Role T-shirtDesigner [] = {"Gabi", "Emil"};
- Role BlouseDesigner [] = {"Clara", "Bale", "Jan"};

Now many roles are defined to execute a process. How can a particular role be selected for a particular data pattern? In a WFMS input data is made available before process execution starts. Based on the features of input data (design features) an appropriate role is selected using the following notions:

- IF(Design.MajorType="Pants" and Design.Stuff="Jeans")
  THEN Performer.Role() = JeansPantsDesigner;
- IF(Design.MajorType="Pants" and Design.Stuff="Dress")
  THEN Performer.Role() = DressPantsDesigner;
- IF(Design.MajorType="Shirt" and Design.MinorType="T-shirt")
  THEN Performer.Role() = T-shirtDesigner;
- IF(Design.MajorType="Shirt" and Design.MinorType="Blouse")
  THEN Performer.Role() = BlouseDesigner;

4 Experimental Results

To investigate the usefulness of our PAPE method experiments we performed on the datasets of two different garment factories; we call them Factory-A and Factory-B. The companies must be kept anonymous. In a collaborative effort of domain experts and workflow designers 26 processes were selected from factory-A and 33 processes from factory-B. Processes which have input data to form data patterns were initially selected and datasets were prepared by integrating process execution data, processes structure data and business benefit data. Due to non-disclosure we cannot provide all details of the dataset in this paper but we report the evaluation of our results.

We are using two graphs to represent these prediction accuracies: the BB-Without-PAPE graph represents prediction accuracy without pattern based agent performance evaluation (data pattern $\rightarrow$ business benefit) and the BB-With-PAPE graph represents prediction accuracy with pattern based agent performance evaluation (data pattern, agent $\rightarrow$ business benefit).

Fig. 4 shows the results of our evaluation using the BB-Without-PAPE and BB-With-PAPE graphs. As shown in Fig. 4 (upper part: Factory-A) there is a significant difference between prediction accuracies of BB-Without-PAPE and BB-With-PAPE graphs for 9 processes; this means that there is a dependency between agent assignment and business benefit. Whereas the prediction accuracies are almost the same for 17 processes; this means that these processes were executed equally well by all agents and no dependency between agent assignment and business benefit were observed. Similarly for factory-B such dependency is observed in 12 out of 33 processes; 21 processes were executed almost equally well (Fig. 4 lower part).

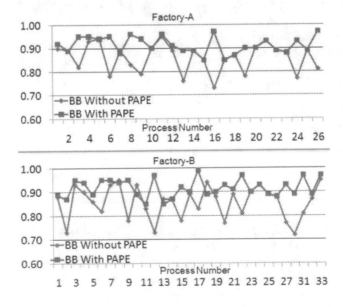

Fig. 4. Measuring the usefulness of the PAPE method

Although pattern based agent performance evaluation is applicable to a comparatively small number of processes it is very valuable for achieving the business benefit objective. We are currently working on the enactment of the PAPE method in these two factories. This means that we aim at the substitution of the current agent assignment rules with the rules learned by our method.

5   Related Work

Process mining techniques are used to support redesign and diagnosis the processes and are proven to be a valuable tool to gain insight into how business processes are being handled within an enterprise [1]? An implicit assumption of process improvement is to benefits the business [10]. But most process mining techniques re-construct a process model from the data recorded in an event log to reflect the behavior of process execution (discovery) [13] [14]. Then this observed behavior is compared with the original process model to detect possible deviations (conformance checking) [13] [15]. Most of the process mining research focuses on the functional and control flow perspective. Relatively small portion of research is found on the organizational perspective.

In [8] [13] van der Aalst et al. constructed the organizational models. These models represent the structure of the organization and the relationships between organizational structures and organizational population: who performs what and how performers are related? In [13], they developed the methods for mining social networks from process logs to analyze relationships between agents involved in the processes. They presented different matrices like *handover of work matrices*, *in-between matrices* and *working together matrices* to express potential relationships between agents. Also, in [8] they developed methods for mining organizational models and analyze relationships among organizational entities and units.

The work in [6] [7] [12] is related to agent assignment. In [7], Ly et al. focused on mining the agent assignments rules from process log using machine learning technique. He combined the organizational model with the audit trail data and learned agent assignment rules. These rules define the profiles of agents capable or eligible of performing an activity but not the profile of agents who execute the process in a way that direct to organizational business benefit.

Similarly in [6], Liu Yingho et al. applied machine learning classifier to workflow event log to learn various kinds of activities each agent undertakes. When a new process is initiated this machine learning classifier suggests a suitable agent who can execute the particular process. But his objective was to reduce the burden of staff assigner for conventional agent assignment rather to automatic agent assignment in WFMS . Moreover the learned classifier predicts only one agent who might be on leave or busy with some other process.

Similarly in [12] S. Rinderle-Ma et al. proposed agent assignment rules mining technique. They compared the mined assignment rules with the pre-defined agent assignment rules in order to detect possible deviations. These deviations detect the security loopholes like offering the process to non authorized agent. The main contribution of this work is to first identify agents working successfully and feedback this information into WFMS to automate future agent assignments.

6   Conclusion and Future Work

In this paper we discussed a method for mining agent assignment rules based on agent performance. Agent performance is evaluated to determine how agents are working successfully for achieving business benefit and this information is feed backed for future agent assignment. Our method evaluate agent performance more precisely using machine learning technique and post-processing technique and suggests agent assignment rules which leads to the success of the performance of a business process as a whole and the agent assignment in detail.

We believe that our method shows some promise for improving the current state of workflow agent assignment strategies. Our future plans include further experiments with substituted agent assignment rules. An empirical study will be performed as an evaluation of adequacy of our PAPE method with the enactment of modified agent assignment rules in these factories. We also investigate to learn agent assignment rules for processes which have input data without having adequate data patterns.

References

1. Rozinat, A., van der Aalst, W.M.P.: Decision Mining in Business Processes. Developmental Review (2006),
   http://prom.win.tue.nl/research/wiki/publications/beta_164
   BPM Center Report BPM-06-10, BPMcenter.org
2. Bussler, C.: Organisationsverwaltung in Workflow-Management-Systemen (in German). Deutscher Universitäts-Verlag (1998)
3. Cao, L., Zhang, C., Yu, P., et al.: Domain-Driven actionable knowledge discovery. IEEE Intelligent Systems 22(4), 78–89 (2007)
4. Witten, I.H., Frank, E.: Data Mining: Practical Machine Learning Tools and Techniques, 2nd edn. Morgan Kaufmann, San Francisco (2005)
5. Jablonski, S., Bussler, C.: Workflow Management: Modeling Concepts, Architecture and Implementation. International Thomson Computer Press (1996)
6. Yingbo, L., Jianmin, W., Jiaguang, S.: A Machine Learning Approach to Semi-Automating Workflow Staff Assignment. In: SAC 2007, Seol, Korea (2007)
7. Ly, L., Rinderle, S., Dadam, P., Reichert, M.: Mining Staff Assignment Rules from Event-Based Data. In: Bussler, C.J., Haller, A. (eds.) BPM 2005. LNCS, vol. 3812, pp. 177–190. Springer, Heidelberg (2006)
8. Song, M., von der Aalst, W.M.P.: Towards Comprehensive Support for Organizational Mining, BETA Working Paper Series, WP 211, Eindhoven University of Technology
9. Moore, C.: Common Mistakes in Workflow Implementations. Giga Information Group, Cambridge (2002)
10. Bannerman, P.L.: Capturing Business Benefits from Process Improvement: Four Fallacies and What to Do About Them. In: BIPI 2008, Leipzig, Germany (2008)
11. Quinlan, R.: C4.5: Programs for Machine Learning. Morgan Kaufmann Publishers, San Mateo (1993)
12. Rinderle-Ma, S., van der Aalst, W.M.P.: Life-Cycle Support for Staff Assignment Rules in Process-Aware Information Systems. BETA Working Paper Series, WP 213, Eindhoven University of Technology, Eindhoven (2007),
    http://wwwis.win.tue.nl/%7Ewvdaalst/publications/p367.pdf

13. van der Aalst, W.M.P., de Medeiros, A.K.A., Weijters, A.J.M.M.: Genetic Process Mining. In: Ciardo, G., Darondeau, P. (eds.) ICATPN 2005. LNCS, vol. 3536, pp. 48–69. Springer, Heidelberg (2005)
14. van der Aalst, W.M.P., van Dongen, B.F.: Discovering Workflow Performance Models from Timed Logs. In: Han, Y., Tai, S., Wikarski, D. (eds.) EDCIS 2002. LNCS, vol. 2480, pp. 45–63. Springer, Heidelberg (2002)
15. van der Aalst, W.M.P.: Business Alignment: Using Process Mining as a Tool for Delta Analysis. In: Grundspenkis, J., Kirikova, M. (eds.) Proceedings of the 5th Workshop on Business Process Modeling, Development and Support (BPMDS 2004). Caise 2004 Workshops, vol. 2, pp. 138–145. Riga Technical University, Latvia (2004)

Concept Learning for Achieving Personalized Ontologies: An Active Learning Approach

Murat Şensoy[1] and Pinar Yolum[2]

[1] Department of Computing Science, University of Aberdeen, AB24 3UE, Aberdeen, UK
m.sensoy@abdn.ac.uk
[2] Department of Computer Engineering, Boğaziçi University, Bebek, 34342, Istanbul, Turkey
pinar.yolum@boun.edu.tr

Abstract. In many multiagent approaches, it is usual to assume the existence of a common ontology among agents. However, in dynamic systems, the existence of such an ontology is unrealistic and its maintenance is cumbersome. Burden of maintaining a common ontology can be alleviated by enabling agents to evolve their ontologies personally. However, with different ontologies, agents are likely to run into communication problems since their vocabularies are different from each other. Therefore, to achieve personalized ontologies, agents must have a means to understand the concepts used by others. Consequently, this paper proposes an approach that enables agents to teach each other concepts from their ontologies using examples. Unlike other concept learning approaches, our approach enables the learner to elicit most informative examples interactively from the teacher. Hence, the learner participates to the learning process actively. We empirically compare the proposed approach with the previous concept learning approaches. Our experiments show that using the proposed approach, agents can learn new concepts successfully and with fewer examples.

1 Introduction

In concept learning approaches, an agent teaches another agent a concept from its ontology by providing positive and negative examples of the concept [1]. Positive examples of the concept are chosen among the instances of the concept, where as the negative examples are chosen among the non-instances. Using these examples, the learning agent tries to learn the concept. In this context, learning a concept means having the ability to correctly classify a new example as an instance or a non-instance of the concept. Once a concept is correctly learned, both the teacher and the learner have the same understanding of the concept. Previous approaches show that agents can successfully teach each other concepts from their ontologies using examples [1,2,3]. Although quality of learning depends on quality of the given examples, previous approaches do not specifically address the problem of how to select the most useful examples for the learner.

Selecting the most useful examples means finding a set of positive and negative examples with which the learner can successfully discriminate the concept from other concepts. Since the focus is on discrimination from other concepts, selection of a concise set of negative examples is more important then selection of positive examples. Unlike positive examples, negative examples should be selected among a diverse set of

L. Cao et al. (Eds.): ADMI 2009, LNCS 5680, pp. 170–182, 2009.

instances that belong to various concepts. However, the number of these instances may not be bounded in many settings. For instance, while the teacher in Example 1 is teaching a *motorcycle* concept to the learner, any instance of any other concept (e.g, Sony TV, Leitz ink, and so on) are all negative examples for the *motorcycle* concept, If negative examples are not chosen intelligently, the number of negative examples required to teach a concept will be high, limiting the usability of the learning in real life.

Example 1. A seller, who has the upper ontology in Figure 1, wants to sell a motorcycle to a consumer. However, the consumer has the lower ontology in Figure 1 and does not know *Motorcycles* concept. Therefore, to convince the consumer for buying a motorcycle, the seller must firstly teach the consumer what a motorcycle is.

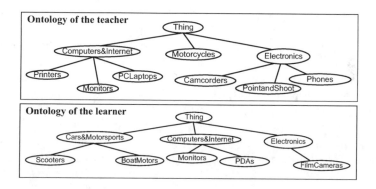

Fig. 1. Ontologies of the teacher (seller) and the learner (buyer)

In current approaches to concept learning, the learner is passive. That is, the training examples are solely chosen by the teacher. However, this assumes that the teacher has an accurate view of what the learner knows, which concepts are confusing for it, and so on. We propose to involve the learner in the learning process by enabling it to interact with the teacher to elicit the most useful examples for its understanding of the concept to be learned. The main contributions of this work are the following:

- Each agent represents its domain knowledge using an ontology and manages this ontology using a network of *experts*. In this way, as in the real life, expertise on a domain is distributed over different domain experts. An expert knows one concept in depth. New concepts are learned by the most related experts.
- Using its expertise and a *semi-supervised learning approach*, an expert first learns the new concept roughly without receiving any negative example from the teacher. Then, it elicits more informative examples from the teacher using *active learning*.
- The comparisons of the proposed approach with existing concept learning approaches (with passive learners) show that our approach can teach an agent a concept better than existing approaches with fewer examples.

2   Representing Knowledge

We consider ontologies that are also taxonomies, such that each node in the taxonomy is a specialization of its parent. Figure 2 shows a taxonomy derived from the product categories of the *Epinions* Website [4]. In machine learning, concepts are usually defined as collections of instances that share certain features [2]. Hence, we assume that there exists a set $F = \{x_1, \ldots, x_n\}$ of features. Then an instance can uniquely be characterized by its values for each of the features, $(x_1, \ldots, x_n)$, where the value of a feature x_i is either 0 or 1. This value denotes whether the instance has the feature or not. The set U denotes the set of all possible instances. Each concept C has a set of instances denoted as $I(C) \in U$, and a set of non-instances denoted as $NI(C) = U - I(C)$. In the problem of instance-based concept learning, a learner agent is given a set of positive examples $PE(C) \in I(C)$ and a set of negative examples $NE(C) \in NI(C)$. Using these examples, the learner trains a classifier to learn C. The concept is assumed to be learned if the probability of misclassification (P_{err}) is less than a threshold. Because the number of examples that can be given is limited, the total number of examples ($|PE(C)| + |NE(C)|$) should be as small as possible without compromising P_{err} much. In this paper, we assume that if two concepts C_A and C_B are sibling concepts in the ontology, then they cannot have a common instance ($I(C_A) \cap I(C_B) = \emptyset$).

In the current instance-based concept learning approaches, one classifier is trained to learn each concept independently [1, 2]. Although the concepts are related through parent-child relationships, their classifiers are regarded as independent of one another. Therefore, if an agent has N different concepts in its ontology, there exist N independently trained classifiers for these concepts. Such approaches require each classifier to learn how to discriminate instances of one concept from those of every other concept in the ontology. Therefore, in order to learn a single concept, the agent uses the whole domain knowledge. In this paper, we envision that the domain knowledge related to an ontology is managed by a set of *experts*, each of which is knowledgeable in a certain concept. By knowledgeable in a concept, we mean that the expert can correctly report which of the concept's subclasses the object belongs to. For example, consider the ontology in Figure 2. An expert on motorcycles (denoted as $A_{Motorcycles}$) can tell us correctly that Burgman 400 is a scooter. Formally, let C be a concept with n child concepts in the concept taxonomy. Each child concept of C in the taxonomy is denoted

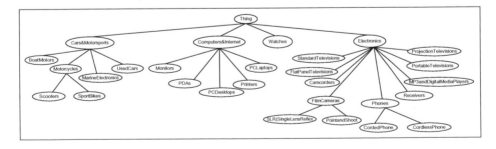

Fig. 2. An ontology from the product categories of Epinions Website

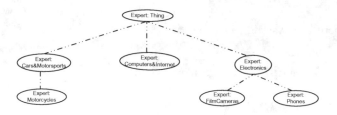

Fig. 3. Experts for the ontology in Figure 2

as C_i, where $i = \{1,\ldots,n\}$. Given that an instance $X \in I(C)$, the expert agent for C, denoted as A, can classify X as an instance of C_j with a confidence $P(C_j|C,X)$, where $j = \{0,\ldots,n\}$. C_0 denotes that the instance belong to C but not any of its child concepts in the taxonomy. Two experts are connected to each other with a communication link if there is a parent-child relationship between the concepts they represent (e.g., A_{Thing} and $A_{Electronics}$). Since each expert has different expertise, in order to train an expert, we do not have to use the whole knowledge of the domain, but we use only the instances of the concept that the agent represents. That is, an expert agent A is given examples from the C_j, where $j = \{0,\ldots,n\}$. A stores these labeled examples and uses them to train a classifier to learn child concepts of C. Given an instance, this expert can categorize the instance into one of $n+1$ classes, where each of the first n classes represents one subconcept of C and the last class represents C's instances that do not belong to any of the subconcepts.

Figure 3 demonstrates the experts for the ontology in Figure 2 and the communication links between them. In the figure, the expert of *Thing* concept can distinguish instances of *Car&Motorsports*, *Computers&Internet*, and *Electronics* concepts, while the expert of *Phones* concept can only distinguish instances of *CordedPhone* and *Cord-lessPhone* concepts. In our approach, experts must cooperate to identify whether an object is an instance of a specific concept or not, because none of these experts have complete expertise on the whole ontology and domain. Initially, the expert that represent the root concept decides whether a given object belongs to any of the subclasses of the concept it represents. If the object belongs to one of the subclasses, then the expert delegates the classification of this object to the expert that represents that subclass. This chain of delegation continues until the expert decides not to delegate the task to any other expert. Example 2 shows how the experts in Figure 3 cooperate.

Example 2. Deciding whether an object o is a CordedPhone requires the computation of $P(Corded\ Phone|o)$. To compute this, first A_{Thing} computes $P(Electronics|Thing,o)$ using its classifier. This is the probability that o is an instance of *Electronics* given that it is an instance of *Thing*. Because every object is an instance of *Thing*, this probability is equal to $P(Electronics|o)$. Then, A_{Thing} passes o to $A_{Electronics}$ together with the computed $P(Electronics|o)$ value. $A_{Electronics}$ computes the probability $P(Phones|Electronics,o)$ and multiplies it with $P(Electronics|o)$ to compute $P(Phones|o)$. Note that $P(A,B|o)$ is equal to $P(B|o)$ if B is a child concept of A. Similarly, $A_{Electronics}$ passes o and the computed $P(Phones|o)$ to A_{Phones}. Lastly, A_{Phones}

computes $P(CordedPhone|o)$ using $P(CordedPhone|Phones, o)$ and $P(Phones|o)$. In this way the object is classified in a cooperative manner.

3   Actively Learning a Concept

Consider Example 1 again, in order to learn *Motorcycles* concept from the teacher's ontology (O^t) and add it into its ontology (O^l), firstly the learner should determine the right place to put *Motorcycles* in O^l. In other words, the learner should determine the parent concept of *Motorcycles* concept in O^l. Parent concept of *Motorcycles* can be defined as the most specific concept subsuming it. For example, *Thing* concept subsumes *Phones* concept, that is $I(Phones) \subset I(Thing)$. However, *Thing* is not the parent of *Phones* concept, because there is a more specific concept *Electronics* that satisfies $I(Phones) \subset I(Electronics)$. Therefore, the parent concept is *Electronics*.

From a machine learning perspective, a concept C_B^l subsumes another concept C_A^t as much as the instances of C_A^t are classified as being an instance of C_B^l. This is actually an estimation of the probability that $I(C_A^t) \subseteq I(C_B^l)$ is true, denoted as $E[P(C_B^l|C_A^t)]$. As usually assumed in the literature, we can assume that the teacher can provide representative set of positive examples, namely $PE(Motorcycles)$. Therefore, using these examples, we can measure how much a concept C_X^l in the learner's ontology subsumes *Motorcycles*. If every motorcycle example $e \in PE(Motorcycles)$ is an instance of C_X^l, namely $P(C_X^l|e) \approx 1.0$, then we can assume that $Motorcycles \subseteq C_X^l$.

We compute $E[P(C_X^l|Motorcycles)]$ by averaging $P(C_X^l|e)$ values for motorcycle examples. Each $P(C_X^l|e)$ value is computed using a set of classifiers. Each classifier may have some uncertainty. This means that even though e is an instance of C_X^l, $P(C_X^l|e)$ may be computed smaller than 1.0. If classifiers are trained enough, this uncertainty should be very small. Therefore, we assume that classifier uncertainty does not exceed a maximum value. We call this the maximum tolerance to the classification uncertainty (ε). If C_X^l subsumes Motorcycles, then the computed $E[P(C_X^l|Motorcycles)]$ value should be greater than $1.0 - \varepsilon$. Otherwise, this average should be significantly smaller than $1.0 - \varepsilon$, because a portion of motorcycle examples should not be classified as an instance of C_X^l, namely $P(C_X^l|e) \approx 0$. This idea is summarized in Equation 1. In order to compute $P(C_X^l|e)$, the learner passes example e to A_{Thing}^l, which is the expert representing *Thing* concept in O^l. Then, e is classified in a cascaded manner by the experts in the hierarchy as explained in Section 2. This way, for each C_X^l and e, $P(C_X^l|e)$ is computed.

$$E\left[P(C_X|C_Y)\right] = \frac{\sum_{e \in PE(C_Y)} [P(C_X|e)]}{|PE(C_Y)|} \tag{1}$$

In our example, the learner determines that the most specific concept C_X^l satisfying $E[P(C_X^l|Motorcycles)] \geq 1.0 - \varepsilon$ in its ontology is *Car&Motorsports*. As a result, the learner believes that $I(Motorcycles) \subseteq I(Car\&Motorsports)$. Confidence of this belief is the value of $E[P(Car\&Motorsports|Motorcycles)]$. The higher this value, the more confident the learner about this belief. Now, there are two possibilities, either *Car&Motorsports* is *semantically equal* or the parent of *Motorcycles* in O^l.

3.1 Determining Semantic Equivalence

Two concepts are semantically equal if their instances are exactly the same. Let C_A^t be a concept from the teacher's ontology and C_B^l be a concept from the learner's ontology. These two concepts are semantically equal if and only if $I(C_A^t) = I(C_B^l)$. In order to believe that C_A^t and C_B^l are equal, the learner must have evidences for $I(C_A^t) \subseteq I(C_B^l)$ and $I(C_B^l) \subseteq I(C_A^t)$. The learner can easily test $E[P(C_B^l|C_A^t)] \geq 1.0 - \varepsilon$ as explained before. This is the evidence for $I(C_A^t) \subseteq I(C_B^l)$. However, testing $E[P(C_A^t|C_B^l)] \geq 1.0 - \varepsilon$ is not possible for the learner, since it does not know C_A^t yet. Therefore, the learner cooperates with the teacher. It sends representative positive examples of C_B^l to the teacher. Then, the teacher computes $E[P(C_A^t|C_B^l)]$ using Equation 1 and sends it back to the learner. The learner decides that C_B^l and C_A^t are semantically equal if $E[P(C_A^t|C_B^l)] \geq 1.0 - \varepsilon$ and then it informs the teacher about this equivalence. In this way, the learner and the teacher map concepts from their ontologies. Confidence of the semantic mapping between C_B^l and C_A^t is $E[P(C_B^l|C_A^t)] \times E[P(C_A^t|C_B^l)]$.

In our example, the learner knows that *Car&Motorsports* subsumes *Motorcycles*. If *Motorcycles* has a semantic equivalent in the ontology of the learner, this must be *Car&Motorsports*, because it is the most specific concept subsuming *Motorcycles*. Therefore, the learner needs to decide whether *Motorcycles* and *Car&Motorsports* are semantically equivalent or not. For this purpose, the learner sends the representative positive examples of *Car&Motorsports* concept in its ontology to the teacher. Then, the teacher computes the degree of subsumption as explained before (see Equation 1) and informs the learner. The learner realizes that *Motorcycles* does not subsume *Car&Motorsports*. Therefore, *Motorcycles* is added to the learner's ontology as a new child concept of *Car&Motorsports*, as explained in the next sections.

3.2 Selection of Examples

In our example, the learner decided that *Motorcycles* should be added to its ontology as a new child of *Car&Motorsports*. If *Car&Motorsports* did not have any child concept before, the learner would create a new expert to represent *Car&Motorsports* concept. However, in our example, *Car&Motorsports* is already represented by an expert ($A_{Cars\&Motor}^l$), because it already has some child concepts. For $A_{Cars\&Motor}^l$, learning *Motorcycles* means discriminating its instances from the other instances of the *Car&Motorsports* concept. For this purpose, $A_{Cars\&Motor}^l$ needs positive and negative examples of *Motorcycles*. It already has a set of positive examples given by the teacher. However, the teacher has not given any negative examples yet. At this point, the main purpose of $A_{Cars\&Motor}^l$ is to elicit the most informative negative and positive examples of *Motorcycles* from the teacher.

We know that a portion of *Car&Motorsports'* instances are not motorcycles. The useful negative examples should come from this portion. The most informative negative examples are the ones that are not obvious. Because the teacher does not have much information about the learner's ontology, it may not provide useful or informative negative examples without interacting with the learner. In this setting, $A_{Cars\&Motor}^l$ interacts with the teacher on the behalf of the learner, because it is an expert on *Car&Motorsports* and its subconcepts.

In previous works, the learners are passive and the teachers define the negative and positive examples to be used during the training of the learners. Because the negative examples are too diverse, the teachers may select the negative examples from the instances of the concepts that are most likely to be confused with the concept to be taught. Afsharchi *et. al.* propose that teachers may give negative examples from the concepts that are similar to the concept to be taught [2], because similar concepts may be confused easily. For the computation of similarity, teachers use the distance between the concepts in their ontologies. For example, a teacher selects the negative examples mostly among the instances of the sibling concepts in its ontology [2]. If the teacher and the learner have similar ontologies, this approach sounds reasonable. However, if their ontologies are highly different, then most of the provided negative examples may be useless for the learner.

In our approach, negative examples are not directly given by the teacher, because the teacher cannot estimate which examples are more useful or informative for the learner. Therefore, initially, $A^l_{Cars\&Motor}$ tries to learn *Motorcycles* roughly without receiving any negative examples from the teacher. For this purpose, it uses a semi-supervised learning approach. Yu *et. al.* [5] and Fung *et. al.* [6] show that, only using positive examples, it is possible to extract some negative examples from the unlabeled examples. In this paper, we follow a similar approach. In our case, $A^l_{Cars\&Motor}$ determines the most obvious negative examples of *Motorcycles* using the positive examples and the unlabeled examples (the known instances of *Car\&Motorsports*). Some of these unlabeled examples should be negative examples of *Motorcycles*, because *Car\&Motorsports* is not semantically equivalent to *Motorcycles*. Motorcycle instances should have some features in common that make them separate from the other instances of *Car\&Motorsports* concept. In order to determine which features are more important for the *Motorcycles* concept, the differences of the feature distributions between the positive examples and the unlabeled examples can be used.

3.3   Significance of Instances

If a feature discriminates motorcycle instances from the other instances of *Car & Motorsports*, this feature should be owned more frequently by the positive examples than the unlabeled examples. For example, both the positive examples of *Motorcycles* and the unlabeled examples have an engine as a feature, because all of these examples are instances of *Car\&Motorsports* concept. Therefore, the feature of having an engine is not significant for *Motorcycles* concept. However, although every positive example of *Motorcycles* has a *saddle*, a considerable portion of unlabeled examples does not contain that feature. This implies that, the feature of having a *saddle* is more significant for characterizing *Motorcycles* concept and may help discriminating motorcycle examples from the rest of the unlabeled examples.

Let f^p_i be the frequency of the feature x_i in positive examples and f^u_i be the frequency of x_i in the unlabeled examples. Then, f^p_i / f^u_i defines how significant x_i is for discriminating instances of *Motorcycles* from the other instances of *Car & Motorsports*. Significance of x_i for the positive examples is denoted as $s(x_i)$. If x_i does not appear more frequently in the positive examples, it is assumed that x_i is not significant ($s(x_i) = 0$).

Otherwise, x_i is assumed to be significant as much as its relative frequency in the positive examples $(s(x_i) = f_i^p / f_i^u)$.

We can estimate how significant an instance I is as a motorcycle example, using the significance of its features. For this purpose, we use Equation 2. This equation states that an instance is significant as a motorcycle example as much as the significance of its features. After computing the significance value for each known instance of *Car&Motorsports* using Equation 2, the obvious negative examples of *Motorcycles* are chosen among the instances that have the least significance values.

$$sig(I) = \sum_{x_i \in I} s(x_i), \; where \; s(x_i) = \{ \begin{array}{l} 0 \quad if \; \frac{f_i^p}{f_i^u} \leq 1.0 \\ \frac{f_i^p}{f_i^u} \; otherwise \end{array} \tag{2}$$

3.4   Putting It Together

Algorithm 1 summarizes how the learner learns a concept from the teacher. Using the positive examples of *Motorcycles* and the obvious negative examples (line 2), $A_{Cars\&Motor}^l$ trains a classifier (line 3). This classifier can roughly discriminate instances of *Motorcycles* from other instances of *Car&Motorsports*. However, the boundary between these classes is not learned precisely, because only the obvious negative examples are used.

Algorithm 1. Algorithm for A_B^l to learn concept C_A^t

1: $PosEx = PE(C_A^t), NegEx = \emptyset$
2: $ObvNegEx = findObvNegEx(PE(C_B^l), PosEx)$
3: $Classifier_A.\text{Train}(PosEx, ObvNegEx)$
4: **while** (true) **do**
5:      samples = getSamples(C_B^l)
6:      labeledSamples=$Classifier_A$.labelSamples(samples)
7:      labeledSamples = getUncertainLabels(labeledSamples)
8:      labeledExamples = consultToTeacher(C_A^t, labeledSamples)
9:      **if** (labeledExamples== $\emptyset$) **then**
10:         return $Classifier_A$
11:     **end if**
12:     $PosEx = PosEx \cup labeledExamples.getPosEx()$
13:     $NegEx = NegEx \cup labeledExamples.getNegEx()$
14:     $Classifier_A.\text{Train}(PosEx, NegEx)$
15: **end while**

Moreover, some of these negative examples can be wrongly chosen. These conditions may seriously affect the performance of the trained classifier. Therefore, the learner $A_{Cars\&Motor}^l$ iteratively elicits more useful negative examples from the teacher and learns this boundary more precisely and correctly (lines 4–15). At each iteration, $A_{Cars\&Motor}^l$ samples instances of *Car&Motorsports* (line 5) and then using the classifier, it labels these sampled instances as instance of *Motorcycles* or not (line 6). Some of these instances are close to the class boundary and so they are classified with low certainty. These labeled instances are chosen to be asked to the teacher (line 7). That is, $A^l_{Cars\&Motor}$ gives the teacher a set of instances and their estimated labels (line 8). Then, the teacher returns corrected labels of the asked instances (line 8). If the teacher does not return any labeled instances, learning phase is ended and the trained classifier is returned (line 9). Otherwise, the returned labeled instances are used by $A^l_{Cars\&Motor}$ to retrain the classifier and a new iteration starts (lines 12-14). In this way, during these iterations, $A^l_{Cars\&Motor}$ elicits the most confusing negative and positive examples of *Motorcycles* for the learner.

Using the procedure above, *Motorcycles* concept is learned correctly. Then, it is placed into the learner's ontology as a new subconcept of *Car&Motorsports*. Lastly, we test whether *Motorcycles* concept subsumes some subconcepts of *Car& Motorsports* or not. If this is the case, concept-subconcept relationships are rearranged.

4   Evaluation

In order to evaluate our approach, we conduct several experiments in online shopping domain. For this purpose, we derive domain knowledge from Epinions [4]. Epinions is a website that contains millions of product reviews and thousands of product descriptions. Products are classified into categories. In order to represent the domain knowledge, we select a subset of these categories and construct the ontology shown in Figure 2. In our experiments, an instance refers to a product item such as *IBM ThinkPad T60*, which is an instance of *PCLaptops* concept. Each product item has a web page in Epinions website and this page contains specification of the product item in English. More than 23,000 different product items are used as instances in this work. We derive a core vocabulary from these specifications and each word in this vocabulary is used as a feature [5].

In our experiments, there is one teacher agent and one learner agent. In the implementation of the agents and the experts, we use JAVA and the C4.5 decision tree classifier of WEKA data mining project [7]. Maximum tolerance to the uncertainty in the classification (ε) is set as 0.05. In each experiment, the teacher and the learner have a different ontology that contains only a subset of the concepts in the Epinions ontology. These ontologies are constructed by randomly choosing 25% of the concepts in the Epinions ontology. Therefore, the teacher and the learner have different ontologies.

We evaluate our approach in three steps. First, we show how successful our approach is in mapping the semantically equivalent concepts in different ontologies. Second, we show how successful our approach is in incrementally learning a new concept from the teacher. Third, we compare our approach with the current approaches and show that our approach learns new concept successfully with fewer negative examples.

Mapping Equivalent Concepts. In our experiments, each concept in the learner's ontology has a semantically equivalent concept in the teacher's ontology. Therefore, the learner finds out the mappings between the concepts in its ontology and the concepts in the ontology of the teacher. Quality of concept mappings only depends on the quality of the knowledge that the teacher and the learner have about the concepts in their ontologies. This knowledge is represented by the related experts. If experts are not trained

Table 1. Average confidence of the mappings between the semantically equal concepts from different ontologies

Concept	Mapped Concept	Confidence
BoatMotors	BoatMotors	0.98
Camcorders	Camcorders	0.98
Cars&Motorsports	Cars&Motorsports	1.0
Computers&Internet	Computers&Internet	0.97
CordedPhone	CordedPhone	0.97
CordlessPhone	CordlessPhone	0.97
Electronics	Electronics	0.99
FilmCameras	FilmCameras	0.97
FlatPanelTelevisions	FlatPanelTelevisions	0.95
MarineElectronics	MarineElectronics	1.0
Monitors	Monitors	1.0
Motorcycles	Motorcycles	0.95
MP3andDigital...	MP3andDigital...	1.0
PCDesktops	PCDesktops	0.98
PCLaptops	PCLaptops	0.99
PDAs	PDAs	0.98
Phones	Phones	1.0
PointandShoot	PointandShoot	0.98
PortableTelevisions	PortableTelevisions	0.96
Printers	Printers	1.0
ProjectionTelevisions	ProjectionTelevisions	0.97
Receivers	Receivers	1.0
Scooters	Scooters	0.93
SLR(SingleLensReflex)	SLR(SingleLensReflex)	0.98
SportBikes	SportBikes	0.98
StandardTelevisions	StandardTelevisions	0.96
UsedCars	UsedCars	1.0
Watches	Watches	1.0

properly or they cannot combine their expertise correctly, either the learner cannot find the correct mappings or the mappings are made with low confidence. In Figure 1, we tabulate the average confidence of the mappings that are concluded by the learner in our experiments. The table shows that our approach can successfully map semantically equivalent concepts to each other with high confidence values. This implies that experts can successfully represent the domain knowledge related to their expertise and successfully combine their expertise.

Incrementally Learning New Concepts. In our approach, a new concept is learned by an expert incrementally on behalf of the learner. At each iteration, the expert labels a set of examples using its existing knowledge about the new concept and consults the teacher. Then, the teacher instructs the expert about the correct labels of these examples. The feedback from the teacher is used to refine and improve the knowledge of the expert about the new concept. Figure 4 shows the performance of our approach at each iteration in terms of the probability of misclassification. In Figure 4, after the first iteration, the expert learns the new concepts roughly. It fails around %12 of its classifications related to this new concept. This error rate is not acceptable for the teacher, so the expert continues with the next iteration. The second iteration results in a considerable progress in the learning performance; the expert classifies more than %95 of sampled instances correctly after the second iteration. The classification error drops to zero at the fifth iteration, which means that the teacher and the learner have exactly the same understanding for this concept. This result shows that the learner has been successful in obtaining useful labeled examples from the teacher as feedback, so that the new concept is learned correctly.

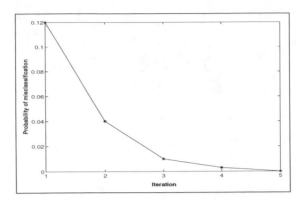

Fig. 4. Misclassification at different iterations

Benchmark Comparisons. We compare our approach with a teacher-driven concept learning approach. This approach represents the current concept learning approaches in the literature. Contrary to the proposed approach, in those approaches, the learner is inactive during the selection of the negative examples [1, 3, 2]. The teacher selects the negative examples using its own ontology and viewpoint. Then, the learner is given positive examples and negative examples of the concept to be taught. Using these examples, the learner trains a classifier to learn the concept. As in the work of Afsharchi *et. al.* [2], in the teacher-driven approach, the teacher selects negative examples of a concept more frequently among the instances of the most similar concepts in its ontology. That is, the more similar a concept is to the concept to be taught, the more probably its instances are selected as negative examples. For the measurement of the similarity, the distance between the concepts within the teacher's ontology is used. One of the main contributions of our approach is learning new concepts with a small number of negative examples. In order to measure how successful our approach is in learning new concept for different number of negative examples, we set up experiments where the teacher is allowed to give or label only a predefined number of negative examples. Then, these examples are given to the learner (as feedback in our approach). After training the learner with these examples, probability of misclassification is computed. Figure 5 compares the results for the teacher-driven approach and the proposed approach.

As seen in Figure 5, the teacher-driven approach requires more negative examples than the proposed approach in order to achieve an acceptable performance. With only five negative examples, the learner that uses the proposed approach fails only on the 12% of its classifications. However, in the same case, the learner using the teacher-driven approach misclassifies an instance with a probability of slightly higher than 0.4. Similarly, with only 35 negative examples, on the average, the proposed approach can learn a concept perfectly, while the teacher-driven approach requires approximately 150 negative examples for the same quality of learning. This result is not surprising, because our approach allows the learner to estimate some negative examples using the unlabeled data. Using these estimated negative examples, the learner can roughly draw a boundary between the instances and non-instances of a concept without receiving

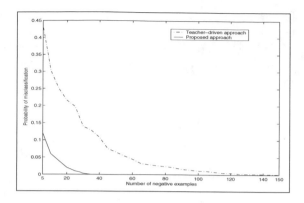

Fig. 5. Misclassification vs. number of negative examples

any negative examples from the teacher. Then, the learner iteratively elicits the most informative negative examples to specify this boundary better.

In the proposed approach, while learning a concept, the learner firstly defines the parent of the concept in its ontology. Then, learning the concept is reduced to discriminating the concept from other child concepts of its parent. This simplifies the original problem, since the concept need not to be differentiated from the other concepts in the ontology. Therefore, our approach requires smaller number of negative examples to learn the concept.

In our experiments, we notice that the learner using teacher-driven approach usually confuses the concepts like *Monitors* and *FlatPanelTelevisions*. These concepts are similar to each other in some way, but they are distant in the ontology of the teacher. Therefore, the teacher regards these concepts as dissimilar. As a result, the teacher gives fewer negative examples from *Monitors* while teaching *FlatPanelTelevisions* and vice versa. For example, in one of our experiments, while learning *FlatPanelTelevisions* concept, the learner using the teacher-driven approach confused *Monitor* and *FlatPanelTelevisions* instances, because only a small portion of the given negative examples was from *Monitors* and these examples were not enough. The learner learned to discriminate *FlatPanelTelevisions* and *Monitors* concepts only after 145 negative examples were given by the teacher and only five of these examples were from *Monitors*.

5   Discussion

This paper develops a framework for instance-based concept learning, where each agent (learner or teacher) represents its domain knowledge with a network of experts that are knowledgeable in different concepts. Using our proposed method, a learner can estimate some negative examples of the concept to be learned and obtain feedback about these negative examples from the teacher to learn the concept accurately. Our experiments show that our approach significantly outperform a teacher-driven approach that represents other instance-based concept learning approaches in the literature by enabling learners to learn a concept with few examples.

Sen and Kar [1] propose an approach for two agents to share a concept. In this approach, the teacher provides predefined positive and negative examples of the concept to be shared to the learner. The learner uses these examples to train a classifier to learn the concept. Sen and Kar show that, using the provided examples, the learner can learn the concept successfully. In Sen and Kar's approach, the learned concepts are not from ontologies and the learner is totally passive.

In order to learn new concepts in a peer-to-peer setting, Afsharchi *et. al.* [2] propose an instance-based approach. In that setting, when an agent confronts an unknown concept, it chooses teacher agents among the other agents in its neighborhood and these teachers teach the concept to the agent by providing positive and negative examples of the concept. Good negative examples are defined as the ones that are similar to the positive examples, because these negative examples are assumed to be easily misclassified. Therefore, a teacher chooses negative examples mostly among the instances of the concepts that are close to the concept to be taught in the concept hierarchy of the teacher (e.g., sibling concepts). After collecting negative and positive examples of the concept, the learner uses a machine learning approach to learn the new concept. In this approach, the learner is not incorporated in the process of choosing negative examples.

Active learning approaches enable learners to engage and involve in the learning process more [8]. In this paper, we propose an approach for concept learning in which learners elicit the most informative training examples from their teachers. To the best of our knowledge, active learning is not used for ontology evolution before. In this paper, we combine instance-based concept learning with the semi-supervised learning and active learning approaches in a novel way. Therefore, our work significantly distinguishes from the literature.

References

1. Sen, S., Kar, P.: Sharing a concept. In: Working Notes of the AAAI 2002 Spring Symposium, pp. 55–60 (2002)
2. Afsharchi, M., Far, B., Denzinger, J.: Ontology guided learning to improve communication among groups of agents. In: Proceedings of AAMAS 2006, pp. 923–930 (2006)
3. Doan, A., Madhaven, J., Dhamankar, R., Domingos, P., Helevy, A.: Learning to match ontologies on the semantic web. VLDB Journal, 303–319 (2003)
4. Epinions web site (1999), http://www.epinions.com
5. Fung, G.P.C., Yu, J.X., Lu, H., Yu, P.S.: Text classification without negative examples revisit. IEEE TKDE 18(1), 6–20 (2006)
6. Yu, H., Han, J., Chang, K.C.C.: PEBL: Web page classification without negative examples. IEEE TKDE 16(1), 70–81 (2004)
7. Witten, I.H., Frank, E.: Data Mining: Practical machine learning tools and techniques. Morgan Kaufmann, San Francisco (2005)
8. Baram, Y., El-Yaniv, R., Luz, K.: Online choice of active learning algorithms. J. Mach. Learn. Res. 5, 255–291 (2004)

The Complex Dynamics of Sponsored Search Markets*

Valentin Robu**, Han La Poutré, and Sander Bohte

CWI, Center for Mathematics and Computer Science
Kruislaan 413, NL-1098 SJ Amsterdam, The Netherlands
{robu, hlp, sbohte}@cwi.nl

Abstract. This paper provides a comprehensive study of the structure and dynamics of online advertising markets, mostly based on techniques from the emergent discipline of complex systems analysis. First, we look at how the display rank of a URL link influences its click frequency, for both sponsored search and organic search. Second, we study the market structure that emerges from these queries, especially the market share distribution of different advertisers. We show that the sponsored search market is highly concentrated, with less than 5% of all advertisers receiving over 2/3 of the clicks in the market. Furthermore, we show that both the number of ad impressions and the number of clicks follow power law distributions of approximately the same coefficient. However, we find this result does not hold when studying the same distribution of clicks per rank position, which shows considerable variance, most likely due to the way advertisers divide their budget on different keywords. Finally, we turn our attention to how such sponsored search data could be used to provide decision support tools for bidding for combinations of keywords. We provide a method to visualize keywords of interest in graphical form, as well as a method to partition these graphs to obtain desirable subsets of search terms.

Keywords: complex systems, sponsored search, keyword advertising, collaborative filtering, power laws, community detection.

1 Introduction

Sponsored search, the payment by advertisers for clicks on text-only ads displayed alongside search engine results, has become an important part of the Web. It represents the main source of revenue for large search engines, such as Google; Yahoo!; and Microsoft's Live.com, and sponsored search is receiving a rapidly increasing share of advertising budgets worldwide. But the problems that arise from sponsored search also present exciting research opportunities, for fields as diverse as economics, artificial intelligence and multi-agent systems.

In the field of multi-agent systems, researchers have been working for some time on topics such as designing automated auction bidding strategies in uncertain and competitive environments (e.g. [4,21]). Another emergent field which studied such topic is agent-based computational economics (ACE) [15], where significant research effort has focused on the dynamics of electronic markets through agent-based simulations.

---

* This work was performed based on a Microft Research "Beyond Search" award. The authors wish to thank Microsoft Research for their support.

** Currently in the Intelligence, Agents, Multimedia Group, University of Southampton, UK.

L. Cao et al. (Eds.): ADMI 2009, LNCS 5680, pp. 183–198, 2009.

One particular topic of research for the ACE community is how order and macro-level market structure can emerge from the micro-level actions of individual users. However, most existing work has been based on simulations, as there are few sources of large-scale, empirical data from real-world automated markets. In this context, empirical data made available from sponsored search provides an excellent opportunity to test the assumptions made in such models in a real market.

In this paper, which is based on large-scale Microsoft sponsored search data, we provide a detailed empirical analysis of such data. To do this, we make use of several techniques derived from computational economics, and especially complex systems theory. Complex systems analysis (which we briefly review below) has been shown to be an excellent tool for analyzing large social, technological and economic systems, including web systems [19,11,6].

1.1 The Data Set

The study provided in this paper is based on a large dataset of sponsored search queries, obtained from the website Live.com[1]. The search data provided consists of two distinct data sets: a set of sponsored search dataset (URLs returned are allocated to advertisers, through an auction mechanism) and an organic search dataset (standard, unbiased web search). The sponsored search data consists of 101,171,081 distinct impressions (i.e. single displays of advertiser links, corresponding to one web query), which in total received 7,822,292 clicks. This sponsored dataset was collected for a roughly 3-month period in the autumn of 2007. The organic search data set consists of 12,251,068 queries, and was collected in a different 3-month interval in 2006 (therefore the two data sets are chronologically disjoint).

It is important to stress that in the results reported in this paper are based mostly on the sponsored search data set[2]. Furthermore, the sponsored search data we had available only provides partial information, in order to protect the privacy of Microsoft Live.com customers and business partners. For example, we have no information about financial issues, such the prices of different keywords, how much different advertisers bid for these keywords, the budgets they allocate etc. Furthermore, while the database provides an anonymized identifier for each user performing a query, this does not allow us to trace individual users for any length of time.

Nevertheless, one can extract a great deal of useful information from the data. For example, the identities of the bidders; for which keyword combinations their ads were shown (i.e. the impressions); for which of these combinations they received a click; the position their sponsored link was in when clicked etc... Insights gained from analyzing this information forms the main topic of this paper.

2   Complex Systems Analysis Applied to the Web and Economics

Complex systems represents an emerging research discipline, at the intersection of diverse fields such as AI, economics, multi-agent simulations, but also physics and

---

[1] This data was kindly provided to us by Microsoft research through "Beyond Search" award.

[2] The only exception is a plot on the distribution number of clicks vs. display rank in Sect. 3, included for comparison reasons.

biology [2]. The general topic of studies in the field of complex systems is how macro-level structure can emerge from individual, micro-level actions performed by a large number of individual agents (such as in an electronic market). For web phenomena, complex systems techniques have been successfully used before to study phenomena such as collaborative tagging [11,10] or the formation of online social groups [1].

One of the phenomena that are indicative to such complex dynamics is the emergence of scale-free distributions, such as power laws. The emergence of power laws in such a system usually indicates that some sort of complex feedback phenomena (e.g. such as a preferential attachement phenomena) is at work. This is usually one of the criteria used for describing the system as "complex" [2,6]. Research in disciplines such as econophysics and computational economics discusses how such power laws can emerge in large-scale economic systems (see [6,19] for a detailed discussion).

2.1 Power Laws: Definition

A *power law* is a relationship between two scalar quantities x and y of the form:

$$y = cx^\alpha \tag{1}$$

where α and c are constants characterizing the given power law. Eq. 1 can also be written as:

$$\log y = \alpha \log x + \log c \tag{2}$$

When written in this form, a fundamental property of power laws becomes apparent; when plotted in log-log space, power laws appear as straight lines. As shown by Newman [19] and others, the main parameter that characterizes a power law is its slope parameter α. (On a log-log scale, the constant parameter c only gives the "vertical shift" of the distribution with respect to the y-axis.). Vertical shift can vary significantly between different phenomena measured (in this case, click distributions), which otherwise follow the same dynamics. Furthermore, since the logarithm is applied to both sides of the equation, the size of the parameter α does not depend on the basis chosen for the logarithm (although the shifting constant c is affected). In the log-log plots shown in this paper, we have chosen the basis of the logarithm to be 2, since we found graphs with this low basis the more graphically intuitive. But, in principle, the same conclusions should hold if we choose the logarithm basis to be, e.g. e or 10.

3 Influence of Display Rank on Clicking Behavior

The first issue that we studied (for both sponsored and organic search data) is how the position that a URL link is displayed in influences its chances of receiving a click. Note that this particular issue has received much attention in existing literature [8]. To briefly explain, Microsoft's Live.com search interface (from which the data was collected), is structured as follows:

- For sponsored search there are up to 8 available slots (positions) in which sponsored URL links can be placed. Three of these positions (ranked as 1-3) appear at the top

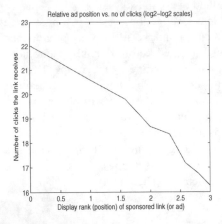
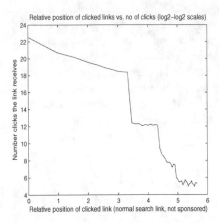

Fig. 1. Distribution of clicks received by a URL, relative to its position on the display, for sponsored and organic search. A(left-side, sponsored search): There are up to 8 sponsored advertiser links displayed: 3 on the top of the page, and 5 in a side bar. B(right, organic search): There are usually 10 positions displayed per page, with multiple result pages appearing as plateaus.

of the page, above the organic search results, but delimited from those by a different background. In addition, the page can display up to 5 additional links in a side bar at the right of the page.

– The "organic" search results are usually returned as 10 URL links/page (a user can opt to change this setting, but very few actually do).

All the sponsored links are allocated based on an auction-like mechanism between the set of interested advertisers (such a display, in any position is called in "impression"). However, the advertisers only pay if their link actually gets clicked - i.e. "pay per click" model. The exact algorithm used by the engine to determine the winners and which advertiser gets which position is a complex mechanism design problem and not all details are made public. However, in general, it depends on such factors as the price the bidder is willing to pay per click, the relevance of the query to her set of terms, and her past performance in terms of "clickthrough rate" (i.e. how often links of that user were clicked in the past, for a given keyword). By contrast, in organic search, returned results are ranked simply based on relevance to the user's query.

3.1   Results on Display Position Bias and Interpretation

Results for the position bias on click distribution are plotted in Fig. 1: part A (left side) for sponsored search and part B (right side) for the organic search. Note that both of these are cumulative distributions: they were obtained by adding the number of clicks for a link in each position, irrespective of the exact context of the queries or links that generated them. Furthermore, both are drawn in the log-log space.

There are two main conclusions to be drawn from these pictures. For the sponsored search results (Fig. 1.A), the distribution across the 8 slots seems to resemble a straight line, with a slope parameter aprox. $\alpha = 2$. However, such a conclusion would be too simplistic: there is, in fact, a difference between the slope between the first 3 positions (up to $log_2 3$, on the horizontal axis), and the last 5 positions. The slope for the first 3 positions is around $\alpha_1 = 1.4$, while for the last 5 is around $\alpha_2 = 2.5$. The most likely reason for this drop comes from the way the Live.com search interface is designed. The first 3 slots for sponsored search links are shown on the top of the page, above the organic search results, while the last 5 are shown in a side bar on the right of the page.

Fig. 1.B corresponds to the same plot for organic search results, the main effect one notices is the presence of several levels (thresholds), corresponding to clicks on different search pages. We stress that, since this is a log-log plot, the drop in attention between subsequent search pages is indeed very large - about two orders of magnitude (i.e. the top-ranked link on the second search page is, on average, about 65 times less likely to be clicked than the last-ranked link on the first page). The distribution of intra-page clicks, however, at least for the first page of results, could be roughly approximated by a power law of coefficient $\alpha = 1.25$.

All this raises of course the question: what do these distributions mean and what kind of user behaviour could account for the emergence of such distributions in sponsored search results? First, we should point out that the fact that we find power law distributions in this context is not completely surprising. Such distributions have been observed in many web and social phenomena (to give just one example, in collaborative tagging systems, in the work by one of the co-authors of this paper [11] and others). In fact, any model of "top to bottom" probabilistic attention behaviour, such as a user scanning the list of results from top to bottom and leaving the site with a certain probability by clicking one of them could give rise to such a distribution. Of course, more fine-grained models of user behavior are needed to explaining click behavior in this context (an example of such a model is [8]). But for now we leave this issue to further research, and we look at the main topic of this paper which is examining the structure of the sponsored search market itself.

4   Market Structure at the Advertiser Level

In this Section, we look at how sponsored search markets are structured, from the perspective of the participants (i.e. advertisers that buy search slots for their URLs). More specifically, we study how relative market shares are distributed across link-based advertisers. We note that in many markets, an often cited rule, also informally attributed to Pareto, is that 20% of participants in a market (e.g. customers in a marketplace) drive 80% of the activity. Here, we call this effect the "market concentration".

In a sponsored search market, the main "commodity" which produces value for market participants (either advertisers and the search engine) is the number of clicks. Therefore, the first thing that we plotted (first, using normal, i.e. non-logarithmic axes) is the cumulative share of different advertisers (see Fig. 2. A. - left side graph). From this

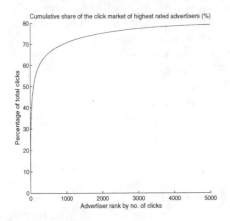 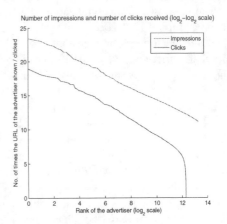

Fig. 2. A (left-side): Cumulative percentage distribution of the number of clicks advertisers in the market receive, wrt. to their rank position, considering the top 5000 advertisers in the market (normal scales). B (right): Log-log scale distributions of the number of impressions, respectively number of clicks, received by the top 10000 advertisers in the market. Note that both distributions follow approximately parallel power laws, but the click distributions levels off in a "long tail" after the first 4000 advertisers, while the impression distribution has a much longer tail (not all appearing in the figure).

graph, one can already see that just the top 500 advertisers get roughly 66% (or about two-thirds) of the total 7.8 million clicks in the available data set[3].

Since in our data, there are *at least* 10000 distinct advertisers (most likely, there are many more, but we only considered the top 10000), this means that a percentage of less than 5% of all advertisers have a two-thirds market share. This suggests that sponsored search markets are indeed very concentrated, perhaps even more so than "traditional" real-world markets.

4.1 Distribution of Impressions vs. Distribution of Clicks for the Top Advertisers

Next, we studied the detailed distribution of the numbers of impressions (i.e. displayed URLs) and clicks on these impressions, for the top 10000 distinct advertisers. Results are shown in Fig. 2.B. (right-hand side graph), using a log-log plot.

The main effect that one can see from Fig. 2.B. is that the distribution of impressions and the distribution for clicks received by the advertisers form two approximately parallel, straight lines in the log-log space (i.e. they are two power laws of approximately the same slope coefficient α). There is one important difference, though, which is the size of the "long tail" of the distribution. The distribution of the number of clicks (lower line), levels off after about 4000-5000 positions. Basically, in data terms, this means that advertisers beyond the top 5000 each receive a negligible number of clicks, at least

---

[3] Note that an advertiser was taken, following the available data, by the domain URL of the sponsored link. This is a reasonable assumption, in this case. For example, Ebay uses many sponsored links to different products. However, using this technique, Ebay is taken as one advertiser, regardless of how many different items its URLs point to.

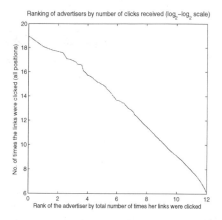
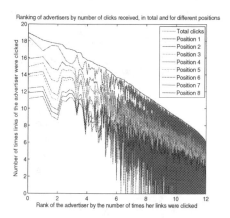

Fig. 3. Distribution of advertiser market share, based on their ordered rank vs. the number of clicks their links receive (log-log scales). The left-hand side plot (part A) gives the total number of clicks an advertiser received for all impressions of her links, regardless of the position they were in. The right-hand side plot (part B) gives the number of clicks received, both in total, but also when her ads were displayed on a specific position on the page (among the 8 ranked slots of the sponsored search interface).

in the dataset we examined. The reason for this may be that their ads almost always appear in the lower display ranks, or simply that they bid on a set of rarely used (or highly specialised) search keywords. By contrast, the distribution of impressions still continues for many more positions (although we only represent the top 10000 distinct advertiser IDs here, as the rest do not play any significant role in the click market).

4.2   Distribution of Market Share Per Display Rank Position

The previous Section examined the power law distributions of the number of clicks each advertiser gets *in aggregate* (i.e. over all display ranks his/her links are shown in). Here, we look how an advertiser's market share distribution is affected when broken down per display rank (an issue we already touched on in Sect. 3).

However, we first make a slight restriction in the number of advertisers we consider. As shown in Sect 4.1 above, there is a power law distribution in the clicks received by the top 4000 advertisers, advertisers ranked beyond this position each receive a negligible number of clicks. Therefore, in this Section, we restrict our attention to the top 4000 advertisers. As these 4000 advertisers receive over 80% of all 7.8 million clicks in the data set (see Fig. 2.A), we do not risk loosing much useful information.

Results are shown in Fig. 3. First, in Fig. 3.A. we show again, more clearly, the power law distribution of the number of clicks for the top 4000 advertisers. Note that this is a "wide" distribution, in the sense that it covers 4000 positions and several orders of magnitude. On the right-hand side graph (Fig. 3.B), we show the same graph, but now, for each advertiser, we also break down the number of clicks received by the position his/her sponsored URL was in when it was clicked.

Surprisingly, perhaps, the smooth power law shape is not followed at the level of the display rank - in fact, for the lower levels the variance becomes so great that the distribution breaks down, at the display rank level. We hypothesize the most likely reason for this variance is the way each individual advertiser does the bidding for the preferred keywords at different points in time, or the way he specifies the way his keyword budget could be used in different periods. For example, some advertisers may have a short-running sale campaign, when they will bid aggresively for the preffered keyword, hence getting the top spot. By contrast, others may prefer to have longer-running ads, even if they don't get the top spot every time. Some anecdotal evidence from online marketing suggests that even just the repeated display of a link of a certain merchant on the screen may count: if a user sees an ad repeatedly in his/her attention space, that may establish the brand as more trustworthy.

In Fig. 3.B, by loking the the top 4 advertisers in this dataset, one can already see that users ranked 2 and 3 utilize a rather different strategy than "the trend" represented by users 1 and 4. While their total numer of clicks does follow, approximately the power law, they seem to get, proportionally speaking, more clicks on the top-ranked slot on the page than the rest. While, in order to preserve the privacy of the data, we cannot mention who these companies are, it does seem that users 2 and 3 are actually "aggregators" of advertising demand. By this, we mean online advertising agencies or engines (or automated services offered by the platform itself) that aggregate demand from different advertisers and do the bidding on their behalf. Apparently, this allows them to capture, proportionally, more often the top slot for the required keyword. Unfortunately, however, we cannot investigate this aspect further, since the dataset provided does not contain any information about bidding, budgets or financial information in general.

In the following and last Section of this paper, we turn our attention to a somewhat different problem: how could we use insights gained from analyzing this query data to provide a bidding decision support for advertisers taking part in this market.

5   Using Click Data to Derive Search Term Recommendations

The previous Sections of this paper used complex systems analysis to provide a high-level examination of the dynamics of sponsored search markets. In this Section, we look at how such query log data could be used to output recommendations to individual advertisers. Such an approach should lead to answers to questions such as: What kind of keyword combinations look most promising to spend one's budget on, such as to attract a maximum number of relevant user clicks? While the previous analysis of power-law formation was done at a macro-level, in this Section we take a more local perspective. That is, we do not consider the set of all possible search terms, but rather a set that is specific to a domain. This is a reasonable model: in practice, most advertisers (which are typically online merchants), are only concerned with a restricted set of keywords which are related to what they are actually trying to sell.

For the analysis in this paper, we have chosen as a domain 50 keywords related to the tourism industry (i.e. online bookings of tickets, travel packages and such). The reason for this is that much of this activity is already fast moving online (e.g. a very substantial proportion of, for example, flight tickets and hotel reservations are now carried out online). Furthermore - and perhaps more important - there are low barriers of entry and the field is not dominated by one major player. This contrasts, for example, other domains, such as the sale of Ipods and accessories, where Apple Stores can be expected to have a dominant position on the clicks in the market.

5.1 Deriving Distances from Co-occurrence in Sponsored Click Logs

Given a large-scale query log, one of the most useful pieces of information it provides is the co-occurence of words in different queries. Much previous work has observed that the fact that two search keywords frequently appear together in the same query gives rise to some implicit semantic distance between them [11].

In this paper, we take a slightly different perspective on this issue, since, in computing the distances, we only use those queries which received at least one sponsored search click for the text ads (i.e. URLs) displayed alongside the results. We argue this is a subtle but very important difference from simply using co-occurence in organic search logs. The fact that queries containing some combination of query words lead to a click on a sponsored URL implies not only a purely semantic distance between those keywords, more important for an advertiser, the fact that users searching on those combinations of keywords have the possible intention of buying things online.

Formally, let $N(T_i, T_j)$ denote the number of times two search terms T_i and T_j appear jointly in the same query, if that query received at least one sponsored search click. Let $N(T_i)$ and $N(T_j)$ denote the same number of queries leading to a click, in which terms T_i, respectively T_j appear in total (regardless of other terms they co-occur with). The cosine similarity distance between terms T_i and T_j is defined as:

$$Sim(T_i, T_j) = \frac{N(T_i, T_j)}{\sqrt{N(T_i) * N(T_j)}} \tag{3}$$

Note that cosine similarity is not the only way to define such a distance, but it is a very promising one in many online application settings, such as shown in previous work by these authors and others [11,16,23,22].

5.2 Constructing Keyword Correlation Graphs

The most intuitive way to represent similarity distances is through a keyword correlation graph. The results from our subset of 50 travel-related terms are shown in Fig. 4. In this graph, the size of each node (representing one query term) is proportional to the absolute frequency of the keyword in all queries in the log. The distances between the nodes are proportional to the similarity distance between each pair of terms, computed Eq. 3, where the whole graph is drawn according to a so called "spring embedder"-type algorithm. In this type of algorithm, edges can be conceived as "springs", whose strength is indirectly proportional to their similarity distance, leading to cluster of edges similar to each other to be shown in the same part of the graph.

There are several commercial and academic packages available to draw such complex networks. The one we think is most suitable - and which was used for graph Fig. 4 - is Pajek (see [3] for a description). Note that not all edges are considered in

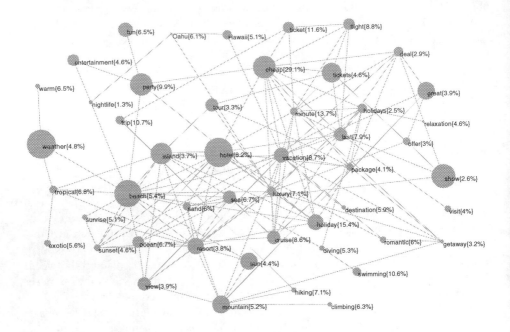

Fig. 4. Visualization of a search term correlation graph, for a set of search terms related to the tourism industry. Each search term is assigned one colored dot. The size of the dots gives its relative weight (in total number of clicks received), while the distances between the dots are obtained through a spring-embedder type algorithm and are proportional to the co-occurrence of the two search terms in a query. Each dot is marked with its success rate (percentage of the total number of impressions associated with that query word that received a click).

the final graph. Even for 50 nodes, there are $\binom{50}{2} = 1225$ possible pairwise similarities (edges), one for each potential keyword pair. Most of these dependencies are, however, spurious (they represent just noise in the data), and our analysis benefits from using only the top fraction, corresponding to the strongest dependencies. In the graph shown in Fig. 4, containing 50 nodes, only the top 150 strongest dependencies were considered in the visualization.

5.3   Graph Correlation Graphs: Results

There are several conclusions that can be drawn from the visualization in Fig. 4 constructed based on the Live.com sponsored search query logs. First, notice that each node was labelled not only with the term or keyword it corresponds to, but also with the aggregate click-through rate (CTR), specific for that keyword. Basicallly, this is the percentage of all the queries that used the term which generated at least one click to a sponsored search URL displayed with that query.

Note that these click-through rates may, at a first glance, seem on the low side: in general only a few percent of all queries actually lead to a click on an sponsored (i.e. advertiser) link. Nevertheless, as a search engine receives millions of queries in a rather short period of time, even a 5%-10% click-through rate can be quite significant. Note that some keywords (such a "cheap") have a higher click-through rate than others. The reason for this may be that people searching for "cheap" things (e.g. cheap airline tickets, cheap holiday packages, hotel rooms etc.) may already have the intention to buy something online, and therefore are more likely to [also] click on sponsored links.

However, the most interesting effect to observe in Fig. 4 are the term clusters that emerge in different parts of the graph, from the application of the spring-embedder visualization algorithm. For example, the leftmost part of the graph has 4 terms related to weather, such as "warm", "tropical" and "exotic". On the top left part of the graph, one can find terms such as "entertainment", "nightlife", "party" and "fun", while very bottom part includes related terms as such "climbing", "hiking" and "mountain". The top-right part includes commercial terms such as: "ticket", "tickets", "flight", "cheap", "last", "minute". The central part of the graph includes terms such a "beach", "sand", "sea", "resort", "ocean", "island" etc. Additionally, pairs of terms one would naturally associate do indeed appear close together, such as "romantic" and "getaway" and "sunset" and "sunrise" and "ocean".

In the following, we discuss an algorithm that can detect such clusters automatically. More precisely, we would like an algorithm that selects combinations of tags that look promising in attracting queries and clicks.

5.4   Automatic Identification of Sets of Keywords

In this Section, we show how keyword graphs could be automatically partitioned into relevant keyword clusters. The technique we use for this purpose is the so called "community detection" algorithm [20], also inspired by complex systems theory. In network or graph-theoretic terms, a community is defined as a subset of nodes that are connected more strongly to each other than to the rest of the network (i.e. a disjoint cluster). If the network analyzed is a social network (i.e. vertexes are people), then "community" has an intuitive interpretation. However, the network-theoretic notion of community detection algorithm is broader, has been succesfully applied to domains such as networks of items on Ebay [12], publications on arXiv, food webs [20] etc.

Community Detection: A Formal Discussion Let the network considered be represented a graph $G = (V, E)$, when $|V| = n$ and $|E| = m$. The community detection problem can be formalized as a partitioning problem, subject to a constraint. Each $v \in V$ must be a assigned to exactly one group (i.e. community or cluster) $C_1, C_2, ... C_{n_C}$, where all clusters are disjoint.

In order to compare which partition is "optimal", the metric used is *modularity*, henceforth denoted by Q. Intuitively, any edge that in a given partition, has both ends in the same cluster contributes to increasing modularity, while any edge that "cuts across" clusters has a negative effect on modularity. Formally, let $e_{ij}, i, j = 1..n_C$ be the fraction of all edge weights in the graph that connect clusters i and j and let $a_i = \frac{1}{2} \sum_j e_{ij}$ be the fraction of the ends of edges in the graph that fall within cluster i. The modularity Q of a graph $|G|$ with respect to a partition C is defined as:

Algorithm 1. *GreedyQ Partitioning*: Given a graph $G = (V,E), |V| = n, |E| = m$ returns partition $< C_1, ...C_{n_C} >$

1. $C_i = \{v_i\}, \forall i = \overline{1,n}$
2. $n_C = n$
3. $\forall i, j, e_{ij}$ initialized as in Eq. 5
4. repeat
5. $\quad < C_i, C_j >= \text{argmax}_{c_i,c_j}(e_{ij} + e_{ji} - 2a_i a_j)$
6. $\quad \Delta Q = \max_{c_i,c_j}(e_{ij} + e_{ji} - 2a_i a_j)$
7. $\quad C_i = C_i \bigcup C_j, C_j = \emptyset$ //*merge C_i and C_j*
8. $\quad n_C = n_C - 1$
9. until $\Delta Q \leq 0$
10. $maxQ = Q(C_1, ..C_{n_C})$

$$Q(G,C) = \sum_i (e_{i,i} - a_i^2) \tag{4}$$

Informally, Q is defined as the fraction of edges in the network that fall within clusters, minus the expected value of the fraction of edges that would fall within the same cluster, if all edges would be assigned using a uniform, random distribution.

As shown in[20], if $Q = 0$, then the chosen partition c shows the same modularity as a random division. A value of Q closer to 1 is an indicator of stronger community structure - in real networks, however, the highest reported value is $Q = 0.75$. In practice, [20] found (based on a wide range of empirical studies) that values of Q above around 0.3 indicate a strong community structure for the given network. In our case, the edges that we considered in the graph (recall that only the strongest 150 edges are considered) have a weight, defined as shown in Eq. 3 above. For the purpose of the clustering algorithm, this weight has to be normalized by the sum of all weights in the system, thus we assign initial values to e_{ij} as:

$$e_{ij} = \frac{1}{\sum_{ij} sim_{ij}} sim_{ij} \tag{5}$$

5.5   The Graph Partitioning Algorithm

The algorithm we use to determine the optimal partition is the "community identification" algorithm described in [20], formally specified as Alg. 1 above. Informally described, the algorithm runs as follows. Initially, each of the vertexes (in our case, each keyword) is assigned to its own individual cluster. Then, at each iteration of the algorithm, two clusters are selected which, if merged, lead to the highest increase in the modularity Q of the partition. As can be seen from lines 5-6 of Alg. 1, because exactly two clusters are merged at each step, it is easy to compute this increase in Q as: $\Delta Q = (e_{ij} + e_{ji} - 2a_i a_j)$ or $\Delta Q = 2 * (e_{ij} - a_i a_j)$ (the value of e_{ij} being symmetric). The algorithm stops when no further increase in Q is possible by further merging.

Cluster 1	Cluster 2	Cluster 3	Cluster 4	Cluster 5	Cluster 6	Cluster 7	Cluster 8	Cluster 9
beach	party	package	weather	getaway	diving	cruise	show	last
luxury	entertainment	vacation	exotic	romantic	swimming	sunrise	tickets	minute
hotel	nightlife	holidays	tropical			sunset	ticket	visit
island	fun	destination	warm				cheap	
resort	Hawaii	deal					flight	
sun	Oahu	tour						
mountain		offer						
ocean		great						
hiking								
climbing								
sea								
sand								
Keywords eliminated to increase modularity: holiday, holidays, relaxation, trip.									

Fig. 5. Optimal partition of the set of travel terms in semantic clusters, when the top 150 edges are considered. The partition was obtained by applying Newman's automated "community detection" algorithm to the graph from Fig 4. This partition has a clustering coefficient Q=0.59.

Note that it is possible to specify another stopping criteria in Alg. 1, line 9, e.g. it is possible to ask the algorithm to return a minimum number of clusters (subsets), by letting the algorithm run until n_C reaches this minimum value. Furthermore, this algorithm is computationally efficient, since it is basically linear in the size of the graph (number of keywords considered), hence it can be applied even to very large datasets.

5.6   Discussion of Graph Partitioning Results

The results from the graph partitioning algorithm, showing the partition maximises the modularity Q for this setting, is shown in Fig. 5. Note that this is not the only possible way to partition this graph - if one would consider a different number of strongest dependencies to begin with (in this case we selected the top 150 edges, for 50 keywords), or a different stopping criteria, one may get a somewhat different result. Furthermore, note that some keywords, which were very general and could fit in several clusters (shown below the figure), were pruned in order to improve modularity, through a separate algorithm not shown here.

Still, the partition results shown in Fig. 5 match well what our intuition would describe as interesting combinations of search terms, for such a setting. There is one large central cluster, of terms that all have reasonably strong relations to each other, and a set of small, marginal clusters on the side. The large cluster in the middle could be further broken by the partition algorithm, but only if we force some other stop criteria than maximum modularity (such as a certain number of distrinct clusters).

The partition in Fig. 5 fits well with what can be graphically observed in Fig. 4: actually, most of the clusters obtained automatically after partition can be identified on different parts of the graph. This does not have to be a one-to-one mapping, however, because in a 2D drawing, the layout of the nodes after "spring embedding" may vary considerably and, furthermore, there are keywords which could fit well into 2 clusters, and were assigned to one as that had a slightly higher modularity.

6   Discussion

6.1   Contribution of the Paper and Related Work

Our work can be seen as related to several other directions of research. Similar techniques to the ones used in this paper have been succesfully applied to analyze large-scale collaborative tagging systems [11] and preference networks for Ebay items [12].

The amount of work which is specifically geared to sponsored search auctions, especially empirical studies, has so far been rather limited (probably not least due to lack of extensive datasets in this field). Much of the previous work, e.g. [8] looks mostly at the bias introduced by a link's display rank on clicking behaviour (such as discussed in Sect. 3 of this paper). Another important direction of work uses existing intuitions about user clicking behaviour to design different allocation mechanisms for this problem - the work of [5] is a good example of this approach. By comparison to our work, the approach taken by [5] studies mostly at mechanism design issues arising from computational advertising, rather than perform an empirical examination of such markets.

One paper that is related in scope to ours, since it also provides an empirical study of search engine advertising markets is [9]. This work takes, however, a different perspective on this problem, also due to the different type of data the authors had available. By contrast to our work, the data that [9] use comes from a single, large-scale advertiser. This means they do get access to more detailed information (including financial one) and can say more about actual bidding behaviour. By comparison, the data available to us for this study does not contain any detailed financial information, but, unlike [9] it allows us to have a global level view of the whole market (from the perspective of the search engine, not just a single advertiser). This provides very important insights about the structure of sponsored search markets.

Finally, there exists previous work that has applied similar co-occurence-based techniques to organic search logs or tagging systems [7,11]. However, our focus in this paper is different: we do not aim to to merely deduce what is the semantic distance between keywords in the general sense, but what kind of combinations of keywords are financially interesting for a sponsored search advertiser to bid on. This is the reason why the size of the nodes and distances computed in Fig. 4 are built using only queries which lead to an actual click on a sponsored ad. Basically, this is equivalent to filtering only the "opinion" (expressed through queries) of the subset of users that are likely to buy something online, rather than all search engine users. To our knowledge, this is the first paper to use sponsored search click data in this way.

6.2   Future Work

This work, being somewhat preliminary, leaves many aspects open to future research. On such aspect would be is the issue of *externalities*: how the presence of links by competing advertisers influences the clickthrough rates of other bidders. As the competition is basically on customers' attention space, externalities play an important role in the efficacity of sponsored search impressions.

Another very interesting topic would be to study the structure of sponsored search markets (in terms of advertiser market share etc.) not only at the global, macro-level, but at the level of individual sets of keywords. In fact, sponsored search can be seen not only as one market, as a network of markets, since most advertisers are interested in (and bid on) a specific set of keywords related to what they are selling. For example, we could apply community detection algorithms to partition not only sets of search keywords, but also sets of bidders (advertisers) interested in those keywords. This should allow us to derive more in-depth insights into the structure of sponsored search.

Finally, exploring how specialized mechanisms such as priced options [13,14,18,17] could improve the allocation of attention space in sponsored search markets represents another promising line for further work.

Acknowledgements

The authors thank Microsoft Research for their support, in the framework of a 'Beyond Search" award. We also wish to thank Nicole Immorlica and Renato Gomes (Nortwestern University) for many useful discussions in the preliminary stages of this work.

References

1. Baldassarri, A., Barrat, A., Cappocci, A., Halpin, H., Lehner, U., Ramasco, J., Robu, V., Taraborelli, D.: The berners-lee hypothesis: Power laws and group structure in flickr, 2008. In: Proc. of Dagstuhl Seminar on Social Web Communities, Dagstuhl DROPS 08391 (2008)
2. Bar-Yam, Y.: The dynamics of complex systems. Westview Press (2003)
3. Batagelj, V., Mrvar, A.: Pajek - A program for large network analysis. Connections 21, 47–57 (1998)
4. Bohte, S.M., Gerding, E., Poutré, J.L.: Market-based recommendation: Agents that compete for consumer attention. ACM Trans. Internet Technology 4(4), 420–448 (2004)
5. Borgs, C., Chayes, J., Immorlica, N., Jain, K., Etesami, O., Mahdian, M.: Dynamics of bid optimization in online advertisement auctions. In: WWW 2007: Proc. 16th Int. Conf. World Wide Web, pp. 531–540. ACM Press, New York (2007)
6. Carter, T.: A short trip through entropy to power laws, 2007. In: Complex Systems Summer School, Santa Fe Institute, NM (2007)
7. Cilibrasi, R.L., Vitanyi, P.M.B.: The google similarity distance. IEEE Trans. on Knowl. and Data Eng. 19(3), 370–383 (2007)
8. Craswell, N., Zoeter, O., Taylor, M., Ramsey, B.: An experimental comparison of click position-bias models. In: Proc of WSDM 2008, pp. 87–94. ACM Press, New York (2008)
9. Ghose, A., Yang, S.: Analyzing search engine advertising: firm behavior and cross-selling in electronic markets. In: WWW 2008: Proc. of the 17th Int. Conf. on World Wide Web, pp. 219–226. ACM Press, New York (2008)
10. Halpin, H., Robu, V., Shepherd, H.: The dynamics and semantics of collaborative tagging. In: Proc. of the 1st Semantic Authoring and Annotation Workshop (SAAW 2006) (2006)
11. Halpin, H., Robu, V., Shepherd, H.: The complex dynamics of collaborative tagging. In: Proc. 16th Int. World Wide Web Conference (WWW 2007), pp. 211–220. ACM, New York (2007)
12. Jin, R.K.-X., Parkes, D.C., Wolfe, P.J.: Analysis of bidding networks in eBay: Aggregate preference identification through community detection. In: Proc. AAAI Workshop on Plan, Activity and Intent Recognition (2007)
13. Juda, A.I., Parkes, D.C.: An options-based method to solve the composability problem in sequential auctions. In: Faratin, P., Rodríguez-Aguilar, J.-A. (eds.) AMEC 2004. LNCS (LNAI), vol. 3435, pp. 44–58. Springer, Heidelberg (2005)

14. Juda, A.I., Parkes, D.C.: The sequential auction problem on ebay: An empirical analysis and a solution. In: Proc. 7th ACM Conf. on Electr. Commerce, pp. 180–189. ACM Press, New York (2006)

15. Judd, K.L., Tesfatsion, L.: Handbook of computational economics II: Agent-Based computational economics. Handbooks in Economics Series. North-Holland, The Netherlands (2005)

16. Linden, G., Smith, B., York, J.: Amazon.com recommendations: item-to-item collaborative filtering. IEEE Internet Computing 7(1), 76–80 (2003)

17. Mous, L., Robu, V., La Poutré, H.: Can priced options solve the exposure problem in sequential auctions? ACM SIGEcom Exchanges 7(2) (2008)

18. Mous, L., Robu, V., La Poutré, H.: Using priced options to solve the exposure problem in sequential auctions. In: Agent-Mediated Electronic Commerce (AMEC 2008). LNCS (LNAI). Springer, Heidelberg (2008) (to appear)

19. Newman, M.: Power laws, pareto distributions and zipf's law. Contemporary Physics 46, 323–351 (2005)

20. Newman, M.E.J.: Fast algorithm for detecting community structure in networks. Phys. Rev. E 69, 66–133 (2004)

21. Robu, V., La Poutré, H.: Designing bidding strategies in sequential auctions for risk averse agents. In: Agent-Mediated Electronic Commerce. LNBIP, vol. 13, pp. 76–89. Springer, Heidelberg (2007)

22. Robu, V., Poutré, H.L.: Learning the structure of utility graphs used in multi-issue negotiation through collaborative filtering. In: Proc. of the Pacific Rim Workshop on Multi-Agent Systems (PRIMA 2005) (2005)

23. Robu, V., Somefun, D., Poutré, J.A.L.: Modeling complex multi-issue negotiations using utility graphs. In: Proc. of the 4th Int. Conf. on Autonomous Agents & Multi Agent Systems (AAMAS), Utrecht, The Netherlands, pp. 280–287. ACM Press, New York (2005)

Author Index